Carbohydrate Chemistry *for* Food Scientists

Roy L. Whistler and James N. BeMiller

eagan® press
St. Paul, Minnesota, USA

Library of Congress Catalog Card Number: 96-78590
International Standard Book Number: 0-913250-92-9

©1997 by the American Association of Cereal Chemists, Inc.
Second printing, 1999

All rights reserved.
No part of this book may be reproduced in any form, including photocopy, microfilm, information storage and retrieval system, computer database or software, or by any other means, including electronic or mechanical, without written permission from the publisher.

Printed in the United States of America on acid-free paper

American Association of Cereal Chemists
3340 Pilot Knob Road
St. Paul, Minnesota 55121-2097 USA

Preface

Food scientists and engineers deal more with carbohydrates than with other food ingredients because of their abundance, low price, food value, and excellent ability to control the physical properties of foods.

This short book on carbohydrate chemistry is designed for use as a classroom text and a handy reference for practical food scientists. Basic principles of carbohydrate chemistry are given to provide an understanding of the chemistry and physical characteristics of monosaccharides, disaccharides, and polysaccharides. With such fundamental information, the food scientist or food engineer can more easily handle simple sugars and polysaccharides.

Some information is provided on the biochemistry and metabolism of carbohydrates to give an understanding of caries formation, of carbohydrate digestion, and of other changes in carbohydrates as they pass through the human gastrointestinal tract. Information is also provided on the function of carbohydrate bulking agents, the use of polysaccharides in the preparation of foods, and the contribution of food gums to food structure and organoleptic properties.

In the presentation of general carbohydrate chemistry, it has been customary to begin with simple sugars and proceed through development of their structures and chemistry before describing oligosaccharides and eventually polysaccharides. Another approach would be to start with polysaccharides and, by description of them, to present the structure and chemistry of simple sugars. Such development would be reasonable for food scientists since the properties and behavior of foods is largely dependent on the prominence of polysaccharides, with less structural and behavioral dependence on simple sugars. However, custom has prevailed, and we have followed the usual route used in teaching carbohydrate chemistry. We have deviated slightly by giving more detail to the polysaccharide chapters so that the advanced student and/or user can skip over the early chapters and still find the later chapters fully explanatory.

Those persons more deeply interested in more detailed information on the character and application of carbohydrates are referred to a number of specialized reference books. Among them are *Industrial Gums—Polysaccharides and Their Derivatives* and *Starch—Chemistry and Technology*, both edited by the authors of this text.

Roy L. Whistler
James N. BeMiller

Contents

1. **Monosaccharides • 1**
   Structures and Nomenclature • Isomerization • Ring Forms • Glycosides

2. **Carbohydrate Reactions • 19**
   Oxidation of the Aldehyde Group and the Anomeric Hydroxyl Group of Aldopyranoses and Aldofuranoses • Reduction of Carbonyl Groups • Oxidation of Nonanomeric Hydroxyl Groups • Esters • Ethers • Cyclic Acetals • Browning

3. **Oligosaccharides • 43**
   Maltose • Lactose • Sucrose • Other Oligosaccharides Related to Sucrose • Oligosaccharides Related to Starch

4. **Polysaccharides • 63**
   Structures • Water Absorption • Solubility and Solution Characteristics • Molecular Associations • Gels • Modification

5. **Behaviors of Polysaccharide Solutions, Dispersions, and Gels • 91**
   Liquids • Gels • Choosing a Thickening or Gelling Agent

6. **Starch • 117**
   Amylopectin • Amylose • Granule Structure • Granule Types • Minor Components of Granules • Gelatinization and Pasting • Retrogradation and Staling • Complexes • Hydrolysis • Modified Food Starch • Manufacture of Starch

7. **Cellulosics • 153**
   Cellulose • Powdered Cellulose • Microcrystalline Cellulose • Carboxymethylcellulose • Methylcelluloses and Hydroxypropylmethylcelluloses

8. **Hemicelluloses • 165**
   Structures and Characteristics • β-Glucan • Corn Fiber Gum • Larch Arabinogalactan

9. **Guar and Locust Bean Gums • 171**
   Sources and Structures • Properties • Uses • Products

10. **Xanthan • 179**
    Structure • Properties • Uses

11. **Carrageenans • 187**
    Structures • Properties • Uses

12. **Alginates • 195**
    Structure • Properties • Uses

13. **Pectins • 203**
    Structures • Properties and Uses

14. **Exudate Gums • 211**
    Gum Arabic • Gum Karaya • Gum Ghatti • Gum Tragacanth

15. **Bulking Agents, Fat Mimetics, and Carbohydrate Nutrition • 217**
    Bulking Agents • Fat Mimetics • Carbohydrate Nutrition

16. **Sweeteners • 225**
    Nutritive Sweeteners • Nonnutritive Sweeteners

Index • 233

Chapter 1

Monosaccharides

Energy reaching the Earth in the form of sunlight is transformed by land and marine plants to sugars, which are used for construction of various plant components and structures. The plant components, in turn, supply food and energy to all other forms of life. Some of the initial photosynthetic carbohydrate material is converted into other organic matter, such as proteins, fats, and lignin. Most of the remaining carbohydrate is converted into polymers of sugars, called *polysaccharides*, which constitute about three-quarters of the dry weight of the biological world and almost 80% of the caloric intake of humans. In the United States, carbohydrates supply almost one-half of the human caloric intake. It is estimated that about 200×10^9 tons of biomass are produced each year by photosynthesis.

Carbohydrates comprise more than 90% of the dry matter of plants. They are common components of foods, both as natural components and as added ingredients, and are also used widely in a variety of other industries. Their use as food ingredients is large in terms of both quantities consumed and variety of applications and products. They are abundant, inexpensive, and can be obtained from a variety of replenishable sources. They occur in diverse structures and degrees of polymerization; thus, they are available in a large variety of molecular sizes, shapes, and solubilities, with varying chemical and physical properties. They are amenable to both chemical and biochemical modification, and both modifications are employed industrially to improve their properties and extend their use. Carbohydrates are safe (nontoxic) and biodegradable. Starch, lactose, and sucrose are digestible by humans and, along with D-glucose and D-fructose, are human energy sources. Although specific carbohydrates are often catabolized only by specific animals, microorganisms in general can utilize any carbohydrate, which allows cattle, horses, sheep, etc. to use forage plant cell wall materials as energy sources.

Structures and Nomenclature

Carbohydrate, a term derived from the German *kohlenhydrate* and the similar French *hydrate de carbone*, expresses the early finding that its general elemental composition is $C_x(H_2O)_y$, which signifies molecules that contain carbon atoms plus hydrogen and oxygen atoms in the same ratio as they occur in water. A later finding showed that carbohydrates are produced by photosynthesis, using carbon dioxide and water as source materials, as indicated by the following unbalanced equation:

$$CO_2 + H_2O \rightarrow Sugar + O_2$$

However, the great majority of natural carbohydrate compounds produced by living organisms do not have this simple empirical formula. Rather, most natural carbohydrate is in the form of oligomers (oligosaccharides, Chapter 3) or polymers (polysaccharides, Chapter 4) of simple and modified sugars. Carbohydrates of lower molecular weight are often obtained by depolymerization of the natural polymers. In this book, we begin with a presentation of the simple sugars and build from there to larger and more complex structures.

Carbohydrates are characterized as having chiral carbon atoms. A *chiral* carbon atom is one that can exist in two different spatial arrangements (configurations). Chiral carbon atoms can be recognized easily; each of their four tetrahedral bonds is connected to a different group. The two different arrangements of the four groups in space (configurations) are nonsuperimposable mirror images of each other. In other words, one is the reflection of the other that we would see in a mirror, with everything that is on the right in one configuration being on the left in the other and vice versa (Fig. 1.1).

The discussion of specific carbohydrate structures begins with D-glucose, the most common, most widely distributed, and most abundant carbohydrate, if all its combined forms are considered. D-Glucose belongs to the class of carbohydrates called monosaccharides, or "simple sugars" in common language. Monosaccharides are carbohydrate molecules that cannot be broken down to simpler carbohydrate molecules by hydrolysis. They are the monomeric building units of oligosaccharides (Chapter 3) and polysaccharides (Chapter 4), which can be broken down to monosaccharides by hydrolysis.

D-Glucose is both a polyalcohol and an aldehyde. It is classified as an *aldose*. The ending *-ose* signifies a sugar. When D-glucose is written in an open or vertical straight-chain fashion, termed an acyclic structure, with the aldehyde group (position 1) at the top and the primary hydroxyl group (position 6) at the bottom, it can be seen that all secondary hydroxyl groups are on carbon atoms that have four different substituents attached to them, making them chiral carbon atoms. Each chiral center has a mirror image (Fig. 1.1). Hence, chiral carbon atoms are not superimposable; just as a person's two hands are mirror images and are not superimposable.[1] Naturally occurring

---

[1] This can be illustrated by pressing the palms of your hands together in front of you. One is the reflection of the other, that is, your thumbs and each of your four fingers oppose each other. Now if you orient your hands in the same direction, for example, with the palms facing away from you, the thumbs are on opposite sides and the hands are different; that is, they are chiral and nonsuperimposable mirror images.

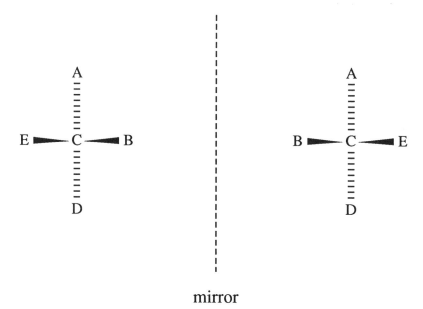

mirror

Fig. 1.1. Chiral carbon atom. A, B, D, and E represent different atoms, functional groups, or other groups of atoms attached to carbon atom C. Wedges indicate chemical bonds projecting outward from the plane of the page; dashes indicate chemical bonds projecting into or below the plane of the page.

glucose is designated as the D form, specifically D-glucose. It has a molecular mirror image, termed the L form, specifically L-glucose.

Since each chiral carbon atom has a mirror image, there are 2^n arrangements of these atoms (n = the number of chiral carbon atoms in the molecule). Glucose has four chiral carbon atoms, namely, C-2, C-3, C-4, and C-5. Thus, in a six-carbon aldose, there are 2^4 or 16 different arrangements of the carbon atoms containing secondary hydroxyl groups, allowing the formation of 16 different six-carbon sugars with an aldehyde end. Eight of these belong to the D series; eight are their mirror images and belong to the L series. All sugars in which the hydroxyl group on the highest-numbered chiral carbon atom (C-5 in this case) is positioned on the right-hand side are arbitrarily termed D sugars and all with the highest numbered chiral atom having its hydroxyl group on the left are designated L sugars.[2] The structures of D- and L-glucose are shown below in their open-chain (acyclic) form (called the Fischer projection) with the carbon atoms numbered in the conventional manner. As is customary, the horizontal lines indicate the covalent chemical bonds to the hydrogen atoms and hydroxyl groups, as in the simplified structure of glucose (on the right). Because the lowermost carbon atom (C-6 in the case of D-glucose) is nonchiral, the relative positions of the atoms and groups attached to it do not need to be designated, and it is usually written as -CH_2OH.

H—C=O	HC=O	C-1
H—C—OH	HCOH	C-2
HO—C—H	HOCH	C-3
H—C—OH	HCOH	C-4
H—C—OH	HCOH	C-5
H—C—OH	CH_2OH	C-6
H	D - Glucose	

---

[2] To make this determination, the carbon chain must be oriented so that each vertical (carbon-to-carbon) bond projects into or below the plane of the page and each horizontal bond projects outward from the plane of the page, even though in solution there is rotation about the vertical bonds that would allow a hydroxyl group to be in any position with respect to the one above (or below) it.

D-Glucose, as its O-6 phosphate ester, is the first sugar of photosynthesis. D-Glucose 6-phosphate (Chapter 2) is present only in small amounts because it is rapidly converted into sucrose (Chapter 4) for transport to various plant parts and into other plant substances. D-Glucose 6-phosphate is also used as an energy source in the plant's metabolism. D-Glucose, the monomeric building unit of cellulose (Chapter 7) and starch (Chapter 6), can be considered to be the most abundant organic compound on Earth.

D-Glucose and other sugars containing an aldehyde group are called *aldoses* (Table 1.1). Like other sugars containing six carbon atoms, D-glucose is a hexose, the most common group of aldoses. The categorical names are often combined; thus, a six-carbon-atom aldehyde sugar is an aldohexose. There are eight D-aldohexoses and eight L-aldohexoses. All these sugars contain the "saccharose" group and are, therefore, monosaccharides. The scheme below shows the saccharose group, where R is a hydrogen atom (as in aldoses) or a -CH$_2$OH group (as in ketoses).

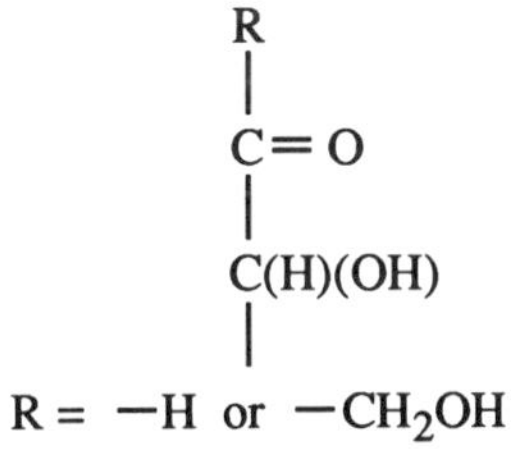

$$\begin{array}{c} R \\ | \\ C=O \\ | \\ C(H)(OH) \\ | \end{array}$$

R = —H or —CH$_2$OH

Saccharose group

There are two aldoses containing three carbon atoms. They are D-glycerose (D-glyceraldehyde) and L-glycerose (L-glyceraldehyde),

TABLE 1.1
Classification of Monosaccharides

Number of Carbon Atoms	Kind of Carbonyl Group	
	Aldehyde	Ketone
3	Triose	Triulose
4	Tetrose	Tetrulose
5	Pentose	Pentulose
6	Hexose	Hexulose
7	Heptose	Heptulose
8	Octose	Octulose
9	Nonose	Nonulose

each possessing but one chiral carbon atom. Aldoses with four carbon atoms, the tetroses, have two chiral carbon atoms; aldoses with five carbon atoms, the pentoses, have three chiral carbon atoms and comprise the second most common group of aldoses. Extending the series above six carbon atoms gives heptoses, octoses, and nonoses (seven, eight, and nine carbon atoms, respectively). Nine carbon atoms is the size limit of naturally occurring sugars; however, only pentoses and hexoses are found in the common carbohydrates of food products and thus are described in this book. Development of the eight D-hexoses from D-glycerose is shown below. The circle represents the aldehyde group; the horizontal lines indicate the location of each hydroxyl group on its chiral carbon atom, and the bottoms of the vertical lines simply indicate the terminal, nonchiral primary alcohol (hydroxyl) group. This shorthand way of indicating monosaccharide structures is called the Rosanoff method (Fig. 1.2). Sugars with names in italics in the family tree below are commonly found in plants, almost exclusively in combined forms. Thus, they are present in our diets in combined forms. Only a small amount of D-glucose in the free monosaccharide form is usually present in natural foods, and it is generally the only free aldose present.

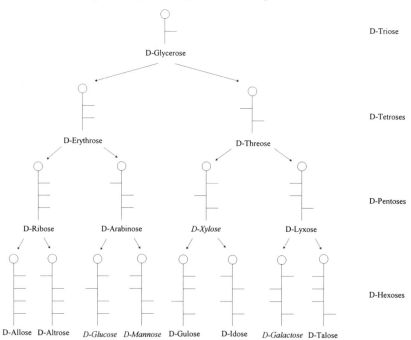

D-Glyceraldehyde (D-glycerose, according to exact carbohydrate nomenclature) is naturally occurring as its O-3 phosphate ester. Most other natural sugars, including the omnipresent D-glucose, have the same configuration of their highest numbered chiral carbon atom. L Sugars are the complete mirror images of the D sugars and are less numerous and less abundant in nature than are the D forms, but they nevertheless have important biochemical roles. The principal L sugar found in foods is L-arabinose, which occurs in a combined form in carbohydrate polymers (polysaccharides, Chapter 4). It is worthwhile to know that the chirality of D-glyceraldehyde, which is the same as that at carbon atom C-5 in D-glucose, occurs in D-amino acids. In contrast, most natural amino acids belong to the L series.

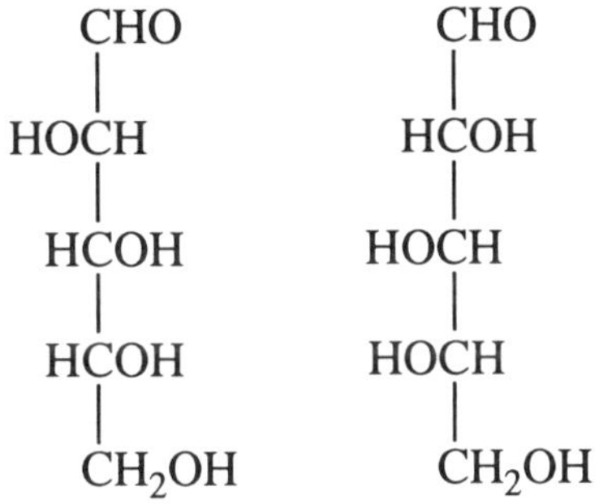

```
      CHO              CHO
       |                |
      HOCH             HCOH
       |                |
      HCOH             HOCH
       |                |
      HCOH             HOCH
       |                |
      CH₂OH            CH₂OH

    D-Arabinose      L-Arabinose
```

To make a likeness of a sugar with models, such as the ball and stick type, one needs to follow two simple rules. The first is to look at only one carbon atom at a time in copying a projected structure such as the Fischer projection or a Rosanoff structure. The second is to keep in mind that all horizontal bonds in the projected structure

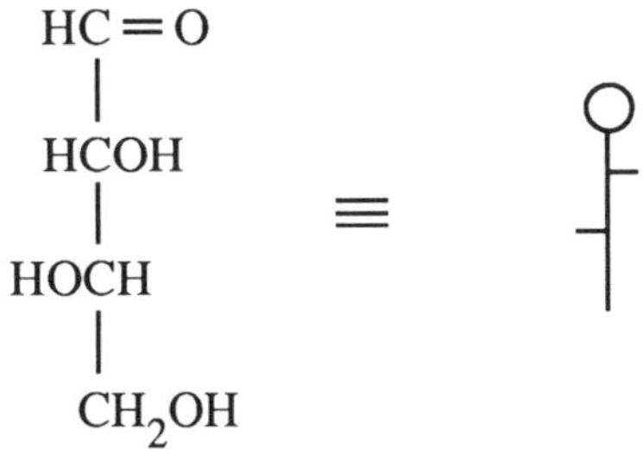

```
      HC=O
       |
      HCOH
       |       ≡      (Rosanoff structure)
      HOCH
       |
      CH₂OH
```

Fig. 1.2. Relation of the Fischer projection to the Rosanoff shorthand projection for L-threose (see below).

are envisioned as protruding forward from the carbon atom, while vertical bonds are envisioned as receding backward, away from it.

So far we have discussed aldoses, in which the carbonyl function of the saccharose group is an aldehyde. In another type of monosaccharide, the carbonyl function of the saccharose group is a ketone group. These sugars are termed *ketoses*. D-Fructose is the prime example of this sugar group. It is one of the two monosaccharide

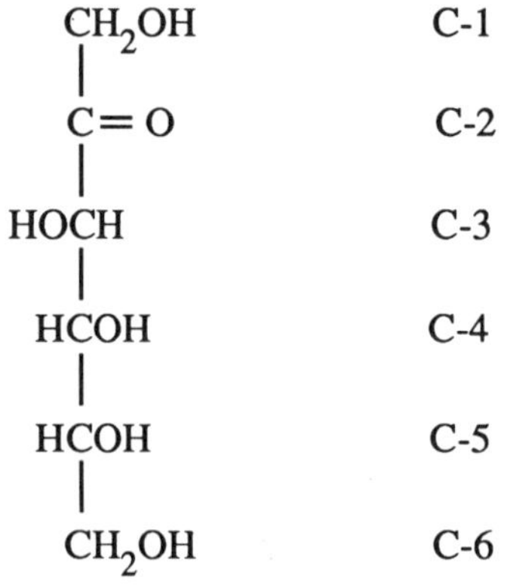

$$
\begin{array}{ll}
CH_2OH & C\text{-}1 \\
| & \\
C=O & C\text{-}2 \\
| & \\
HOCH & C\text{-}3 \\
| & \\
HCOH & C\text{-}4 \\
| & \\
HCOH & C\text{-}5 \\
| & \\
CH_2OH & C\text{-}6
\end{array}
$$

D-Fructose

units of the disaccharide sucrose (the other being D-glucose) (Chapter 3), and it makes up about 55% of most commercial high-fructose corn syrup (Chapter 6) and about 40% of honey. The carbonyl group may be positioned on carbon atoms other than C-2, but such ketoses are uncommon in nature.

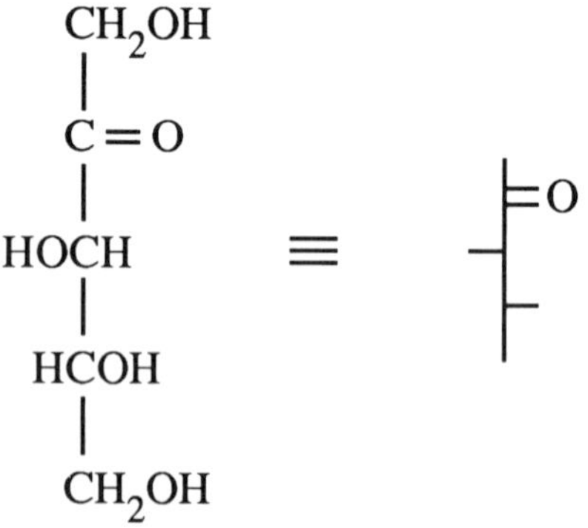

Fig. 1.3. Rosanoff projection of a ketopentose (D-*threo*-pentulose, "D-xylulose"), showing the configurations of the two chiral carbon atoms.

D-Fructose has only three chiral carbon atoms, C-3, C-4, and C-5. Thus, there are but 2^3 or 8 D-ketohexoses. D-Fructose is the principal commercial ketose and the only one of importance in foods. The various ketotetroses, -pentoses (Fig. 1.3), and -hexoses are related to the nonchiral triulose dihydroxyacetone.

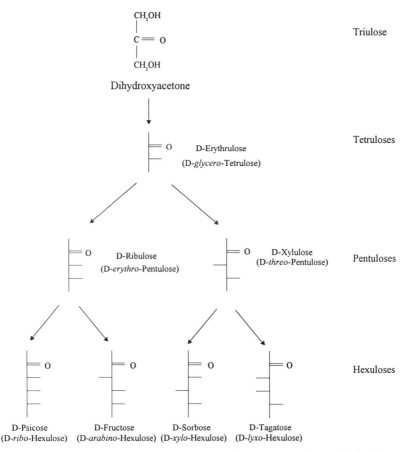

The suffix designating a ketose in systematic carbohydrate nomenclature is *-ulose* (Table 1.1). The generic term for any aldose is *glycose*, and for any ketose, *glycosulose*.

Isomerization

Simple aldoses and ketoses containing the same number of carbon atoms are isomers of each other; that is, a hexose and a hexulose both have the empirical formula $C_6H_{12}O_6$. Isomerization of mono-

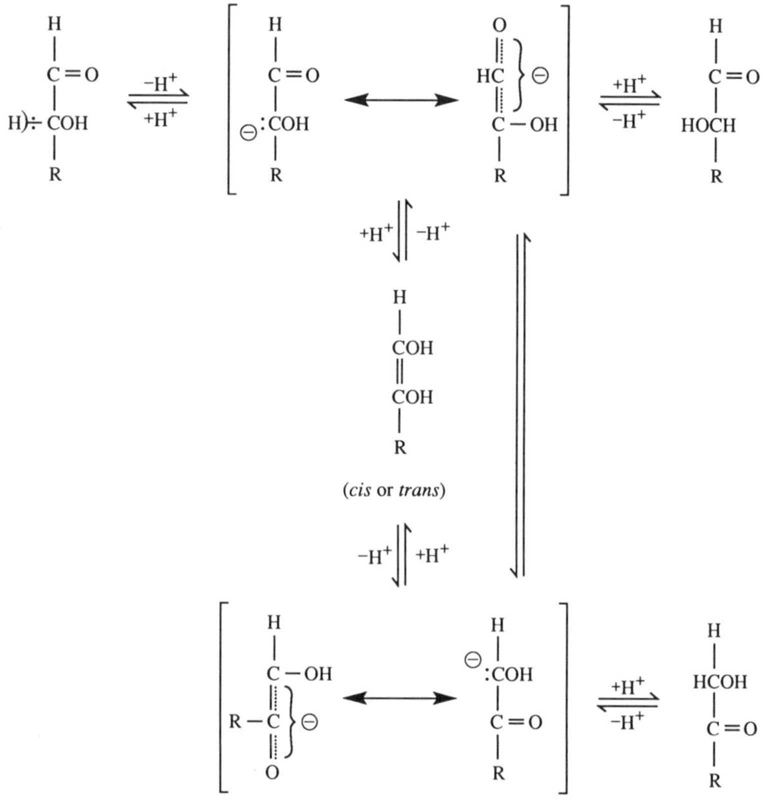

Fig. 1.4. Monosaccharide isomerization.

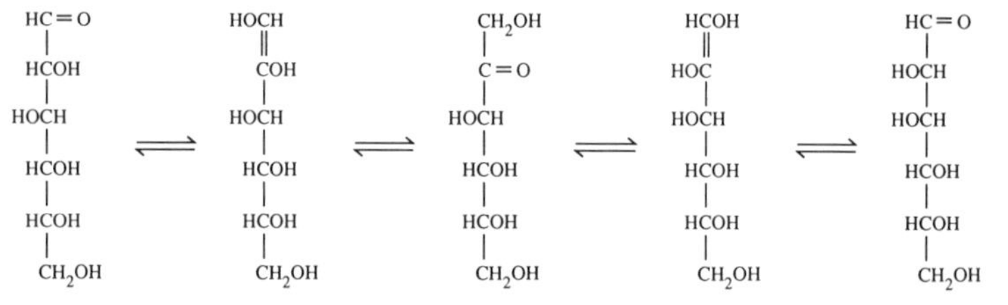

Fig. 1.5. Interrelationship of D-glucose, D-mannose, and D-fructose via isomerization.

saccharides involves both the carbonyl group and the adjacent hydroxyl group. By this reaction, an aldose is converted into another aldose (with the opposite configuration of C-2) and the corresponding ketose, and a ketose is converted into the corresponding two aldoses (Fig. 1.4). Therefore, by isomerization, D-glucose, D-mannose, and D-fructose can be interconverted (Fig. 1.5). Isomerization can be catalyzed by either a base or an enzyme (see the section on glucose and fructose production in Chapter 6).

Ring Forms

Carbonyl groups of aldehydes are reactive and easily undergo nucleophilic attack by the unshared electrons of the oxygen atom of a hydroxyl group to produce a hemiacetal. The hydroxyl group of a hemiacetal can react further (by condensation) with a hydroxyl group of an alcohol to produce an acetal and water. (In the equation below, the italicized atoms are those that are removed to form the water molecule.) By similar reactions, a carbonyl group of a ketone produces an acetal, sometimes specifically designated a *ketal*. Formation of an acetal with methanol is used as an example below.

$$\underset{R}{\overset{CH_3OH}{\underset{+}{H-C=O}}} \rightleftharpoons \underset{R}{\overset{OCH_3}{H-C-OH}} + HOCH_3 \rightleftharpoons \underset{R}{\overset{OCH_3}{H-C-OCH_3}} + H_2O$$

$$\text{Hemiacetal} \qquad\qquad\qquad \text{Acetal}$$

Hemiacetal formation can occur within the same aldose or ketose sugar molecule wherein the carbonyl function reacts with one of its own properly positioned hydroxyl (alcohol) groups, as illustrated with D-glucose, laid coiled on its side, in Figure 1.6. The resulting six-membered sugar ring is called a *pyranose ring*. Notice that, for the oxygen atom of the hydroxyl group at C-5 to react to form the ring, C-5 must rotate to bring its oxygen atom upward. This rotation brings the hydroxymethyl group (C-6) to a position above the ring. The representation of the D-glucopyranose ring used in Figure 1.6 is termed a *Haworth projection*.

To avoid clutter in writing the ring structures, common conventions are adopted wherein ring carbon atoms are indicated by angles in the ring and hydrogen atoms attached to carbon atoms are eliminated altogether. A mixture of chiral forms is indicated by a wavy line.

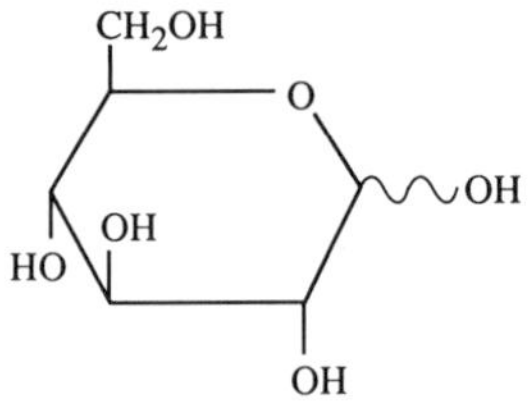

D-Glucopyranose

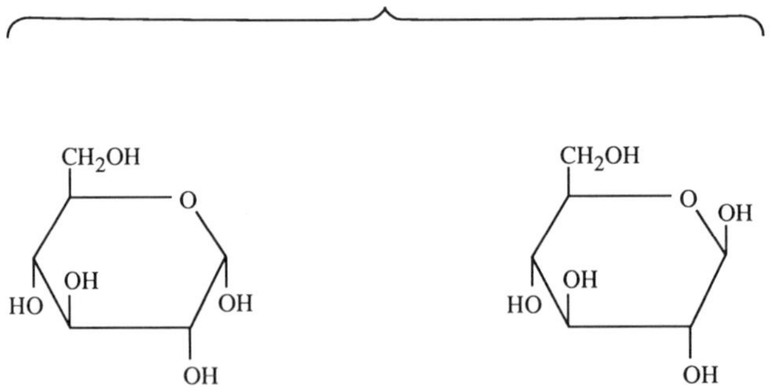

α-D-Glucopyranose   β-D-Glucopyranose

Sugars occur less frequently in five-membered (furanose) rings (Fig. 1.7). The six-membered pyranose ring is more stable than the furanose ring because of bond distances and angles closer to those for a tetrahedral carbon atom in an acyclic structure and less steric interaction of hydroxyl groups. Although the more-strained five-membered (furanose) and seven-membered (septanose) rings occur, the amounts of these higher-energy forms are limited.

When the carbon atom of the carbonyl group is involved in ring formation, leading to hemiacetal (pyranose or furanose ring) devel-

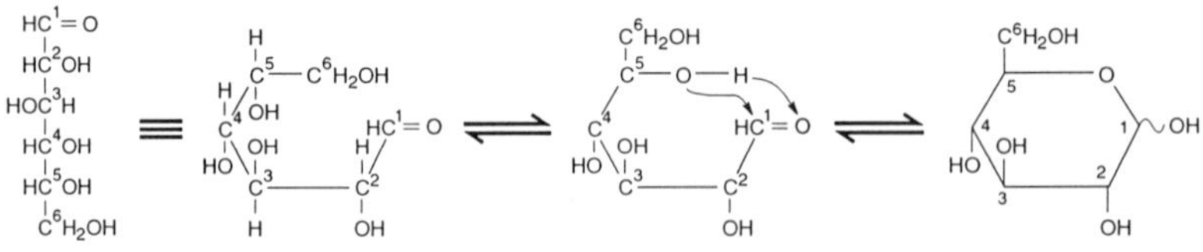

D-Glucose
(Fischer projection)

D-Glucopyranose
(Haworth projection)

Fig. 1.6. Relationship of the acyclic and pyranose ring (Haworth projection) structures of D-glucose.

opment, it becomes chiral because it then has four different groups attached to it. With D sugars, the configuration that has the hydroxyl group located below the ring is the alpha form. For example, α-D-glucopyranose is D-glucose in the pyranose (six-membered) ring form with the hydroxyl group of the new chiral carbon atom, C-1,[3] in the alpha position. When the newly formed hydroxyl group at C-1 is above the ring, it is in the beta position, and the structure is termed β-D-glucopyranose (see next page). This designation holds for all D sugars. For sugars in the L series, the opposite is true; that is, the anomeric hydroxyl group is up in the alpha anomer and down in the beta anomer,[4] so that α-D-glucopyranose and α-L-glucopyranose are mirror images of one another.

The six-membered pyranose ring distorts the normal carbon and oxygen atom bond angles less than do rings of other sizes. The strain is further lessened when the bulky hydroxyl groups are separated maximally from each other by a ring conformation (shape) that arranges the greatest number of them in equatorial rather than axial positions (Fig. 1.8). The pyranose ring can be pictured as a flattened sphere in which one part of the edge is bent slightly up and the edge on the opposite side (when looking down on it) is bent slightly down. The equatorial positions are those that project outward around the perimeter (equator) of the structure, and the axial positions are those that project above and below the flattened sides. This can be

---

[3] The new chiral carbon atom, which was the carbonyl carbon atom in the acyclic form, is termed the *anomeric* carbon atom.

[4] The α and β ring forms of a sugar are known as *anomers* of each other. The two anomers comprise an anomeric pair.

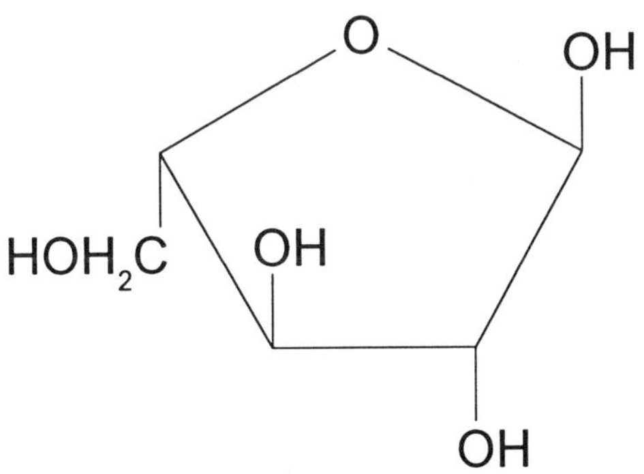

Fig. 1.7. α-L-Arabinose in the furanose ring form.

seen in the space-filling structure of β-D-glucopyranose (Fig. 1.9). The equatorial position for larger groups like hydroxyl groups is energetically favored, and rotation of carbon atoms takes place on their connecting bonds to swivel the bulky groups to equatorial positions in so far as possible. This causes the ring to assume a non-flat conformation.

With ß-D-glucopyranose, for example, C-2, C-3, C-5 and the ring oxygen atom remain in a plane, but C-4 is raised slightly above the plane and C-1 is positioned slightly below the plane, as in Figures 1.8–1.10, giving the ring somewhat of a chair shape. This conformation is designated $^4C_1$. The notation C indicates that the ring is chair-shaped; the superscript and subscript numbers indicate

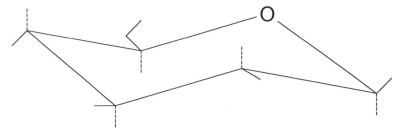

Fig. 1.8. A pyranose ring showing the equatorial (solid line) and axial (dashed line) bond positions.

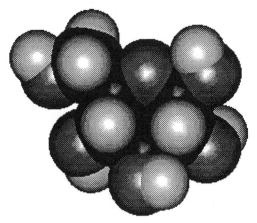

Fig. 1.9. Top view of a space-filling model of β-D-glucopyranose in the $^4C_1$ conformation. (The darkest atoms are carbon atoms; the lightest atoms are hydrogen atoms; and the medium-shaded atoms are oxygen atoms.) The internal energy of the structure has been minimized by computer modeling. All bulky groups (-OH and -CH$_2$OH groups) are in equatorial positions, and only two hydrogen atoms are in axial positions.

that C-4 is above the plane and C-1 below the plane of the ring. There are two chair forms. The second, $^1C_4$, has all the axial and equatorial groups reversed and, therefore, has the bulky groups of D-glucopyranose in the axial position, causing steric, nonbonding interference. Since it is a higher energy form, little of D-glucopyranose exists in the $^1C_4$ conformation.

As noted, ß-D-glucopyranose has all of its hydroxyl groups in the equatorial arrangement. Each is located either slightly above or slightly below the true equatorial position. They alternate in an up-and-down arrangement, with the hydroxyl group at C-1 positioned slightly up and that on C-2 positioned slightly down and continuing with an up-and-down arrangement. The bulky hydroxymethyl group, C-6 in hexoses, is almost always in a sterically free equatorial position.

In addition to the two chair forms, other conformations can exist, depending on the configuration of the sugar and any groups that may be attached to it. Thus, there are six boat conformations and six

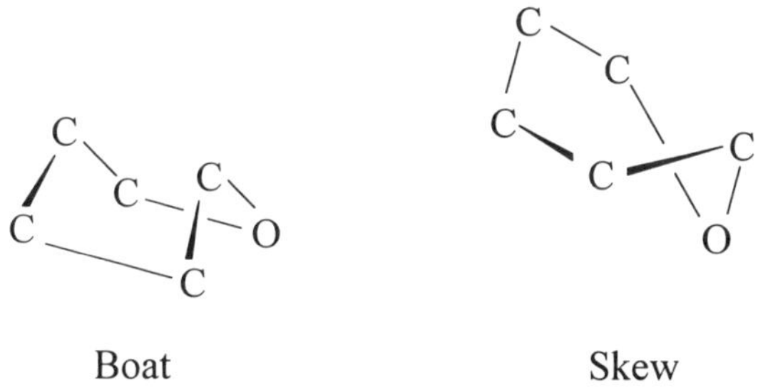

Boat   Skew

skew (twisted boat) shapes possible for the pyranose ring. Few molecules assume these higher-energy forms, although, in solution, they may develop momentarily.

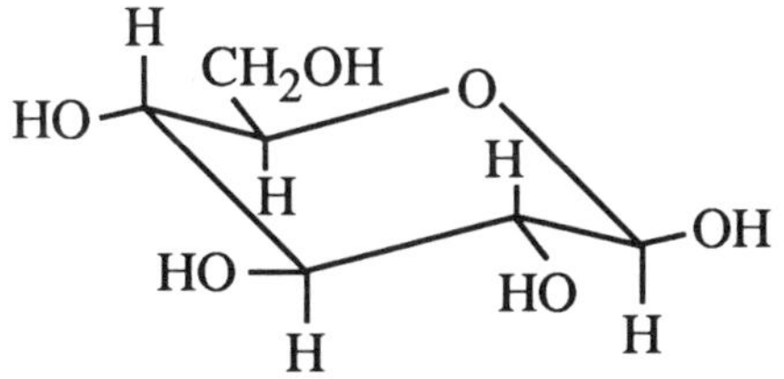

Fig. 1.10. β-D-Glucopyranose in the $^4C_1$ conformation as drawn in structural formulas. All bulky groups are in equatorial positions, all hydrogen atoms in axial positions.

Six-membered sugar rings are quite stable if bulky groups such as hydroxyl groups and the hydroxymethyl group are in equatorial positions. Thus, ß-D-glucopyranose dissolves in water to give a rapidly equilibrating mixture containing the open chain form and its five-, six-, and seven-membered ring forms. At room temperature, the six-membered (pyranose) ring forms predominate, followed by the five-membered (furanose) ring forms and a trace of the seven-membered ring forms; the anomeric arrangement of each ring may be alpha or beta. The open-chain, aldehydo form constitutes only about 0.003% of the total forms.

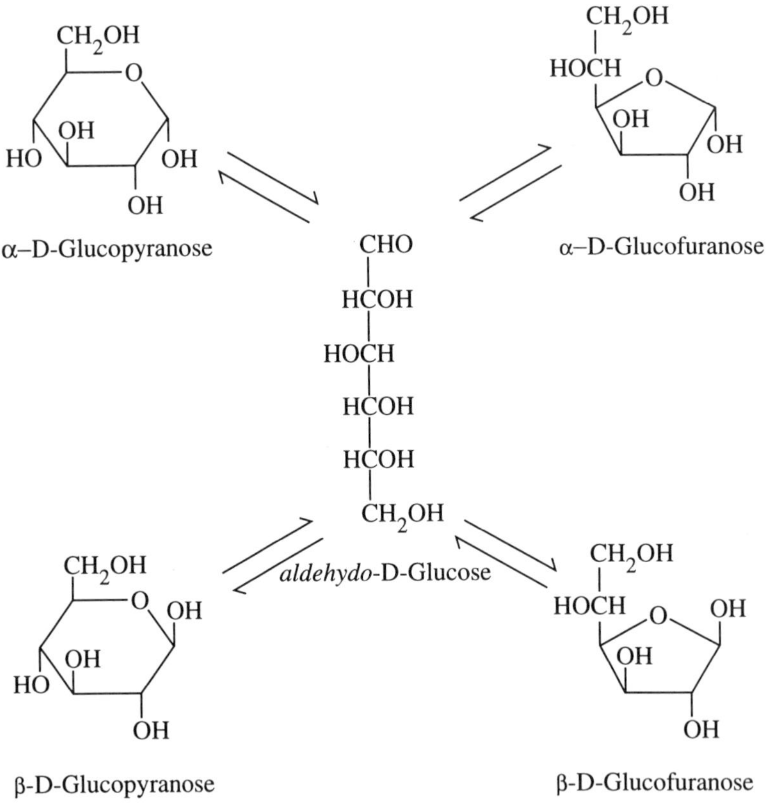

α–D-Glucopyranose   aldehydo-D-Glucose   α–D-Glucofuranose

β-D-Glucopyranose   β-D-Glucofuranose

Glycosides

When the hemiacetal form of sugars reacts with an alcohol (under anhydrous conditions and in the presence of H⁺ as a catalyst) to produce a full acetal, the product is a *glycoside*. The acetal linkage at the anomeric carbon atom is indicated by the *-ide* ending. In the case of D-glucose reacting with methanol, the product is mainly methyl α-D-glucopyranoside, with less methyl ß-D-glucopyranoside. The two anomeric forms of the five-membered-ring furanosides are also formed, but being higher-energy structures, their stability is lower than that of the six-membered pyranoside rings. Although formed initially in substantial amounts, they reorganize under the hot acidic conditions into more stable forms and are present at equilibrium in comparatively low quantities. The methyl group in this case, and any other group bonded to a sugar to make a glycoside, is termed an *aglycon*.

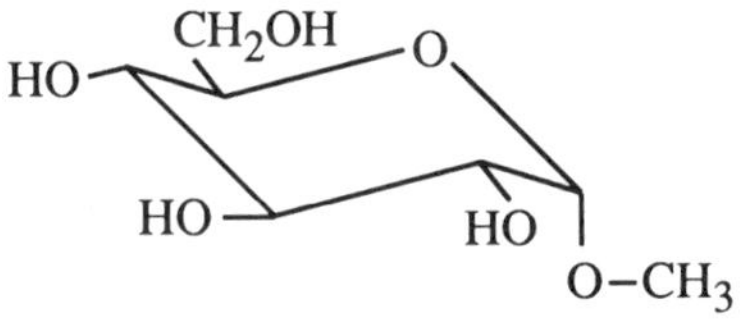

Methyl α-D-glucopyranoside

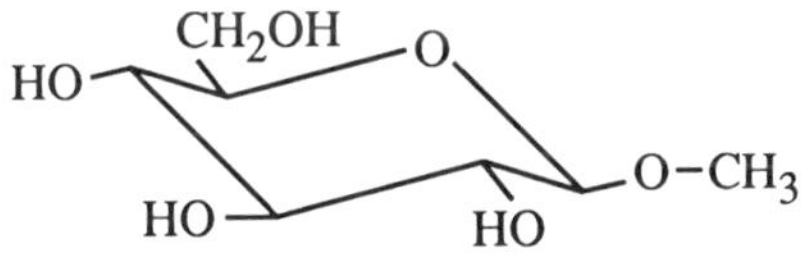

Methyl β-D-glucopyranoside

The α-D-glucopyranoside is the lowest energy form of common D-glucopyranoside and hence the most stable structure in water, even though this places the aglycon group in an axial position. The reason for this is twofold. The dipoles of the nonbonding electrons of the ring oxygen atom and the oxygen atom at C-1 are in a relation in which one bisects the other in the α-D form; this relation is more stable than the parallel relation that occurs in the equatorial ß-D- form. The second, and perhaps stronger, reason is that one pair of the nonbonding electron orbitals of the ring oxygen atom can overlap favorably with the lone pair of electron orbitals on the back lobe of the C-1 oxygen atom in the α-D form to give greater stability, a larger bond angle to the C-O bond, and greater bond length due to the weakening effect of the aglycon group.

Chapter 2

Carbohydrate Reactions

All carbohydrate molecules have hydroxyl groups available for reaction. Simple monosaccharide and most oligo- and polysaccharide (Chapters 3 and 4) molecules also have carbonyl groups available for reaction. Reactions of the carbonyl and hydroxyl groups of carbohydrates are summarized in Table 2.1. Formation of pyranose and furanose rings (cyclic hemiacetals) and glycosides (acetals) of monosaccharides was covered in Chapter 1. Other general reactions are described in this chapter.

Oxidation of the Aldehyde Group and the Anomeric Hydroxyl Group of Aldopyranoses and Aldofuranoses

Aldonic Acids

Aldoses are readily oxidized to aldonic (glyconic) acids. The reaction is commonly used for quantitative determination of sugars and for the manufacture of acids, such as commercial D-gluconic acid.

A qualitative method for determining the presence of aldoses—in fact, any aldehyde—is the Tollens silver mirror test. The Tollens reagent is a basic solution of silver ammonia complex, $Ag(NH_3)_2^+$. The oxidizing agent, silver ion, which converts the aldehyde (aldose) to a carboxylic acid salt (aldonate), is reduced to silver metal, $Ag(s)$, producing a silver mirror coating on the inside of a test tube. The reaction is:

$$Ag(NH_3)_2^+ + R\text{-}\underset{\underset{H}{|}}{C}\text{=}O + 3OH^- \longrightarrow 2Ag(s) + R\text{-}\overset{\overset{O}{\|}}{C}\text{-}O^- + 4NH_3 + 2H_2O$$

One of the earliest methods for quantitative measurement of sugars employed Fehling solution, which is an alkaline solution of a copper(II) salt that oxidizes an aldose to an aldonate and in the process is reduced to copper(I), which precipitates as the brick-red oxide Cu_2O, as shown in the following reaction:

$$2Cu(OH)_2 + R-\underset{H}{\overset{|}{C}}=O \longrightarrow R-\underset{OH}{\overset{O}{\|}}-C + Cu_2O + H_2O$$

Variations, the Nelson-Somogyi and Benedict reagents, are still used for determining amounts of reducing sugars in foods and other biological materials.

Aldoses are called *reducing sugars* because they effect reduction of an agent, such as silver or copper(II) ions, that will oxidize their aldehyde group to a carboxylate group. Ketoses are also termed *reducing sugars* because, under the alkaline conditions of the Fehling test, ketoses are isomerized to aldoses before oxidation (Chapter 1). Benedict reagent, which is not alkaline, will react with aldehydes (aldoses), but not with ketones (ketoses).

TABLE 2.1
Important Reactions of Carbohydrate Molecules

Group Modified	Reactions[a]
Carbonyl group (alone)	1. Oxidation to a carboxylic acid group*
	2. Reduction to a hydroxyl group*
	3. Cyanohydrin reaction and other additions of nucleophiles
Hydroxyl groups	1. Ester formation*
	2. Ether formation*
	3. Cyclic acetal formation*
	4. Oxidation to a carbonyl group*
	5. Reduction to a deoxy carbon atom*
	6. Replacement with amino,* thiol, and halogeno groups
Both carbonyl and hydroxyl groups	1. Formation of cyclic hemiacetals (pyranose and furanose ring forms)*
	2. Formation of acetals (glycosides)*
	3. Aldose ↔ ketose isomerizations*
Anomeric hydroxyl group	1. Formation of glycosyl halides
	2. Oxidation to lactones*

[a] Asterisks indicate common biochemical reactions.

These methods are not stochiometric, and each requires careful control to give quantitative results, even with a standard curve. A stochiometric method, oxidation with hypoiodite anion, IO⁻, at pH 9.5 is used. Hypochlorite and hypobromite similarly oxidize aldoses to aldonic acids but also can oxidize secondary alcohol groups, probably, in the case of an alkaline chlorine solution, by forming the hypochlorite ester as an intermediate.

$$\text{R-CH=O} + I_2 + 3\text{NaOH} \longrightarrow \text{R-C(=O)-O}^-\text{Na}^+ + 2\text{NaI} + 2\text{H}_2\text{O}$$

Glucose Oxidase

A simple and specific method for quantitative oxidation of D-glucose to D-gluconic acid uses the enzyme glucose oxidase, the initial product being the 1,5-lactone (δ-lactone) of the acid. The reaction is commonly employed to measure the amount of D-glucose in foods and biologicals, including the D-glucose level of blood. D-Gluconic acid is a natural constituent of fruit juices and honey.

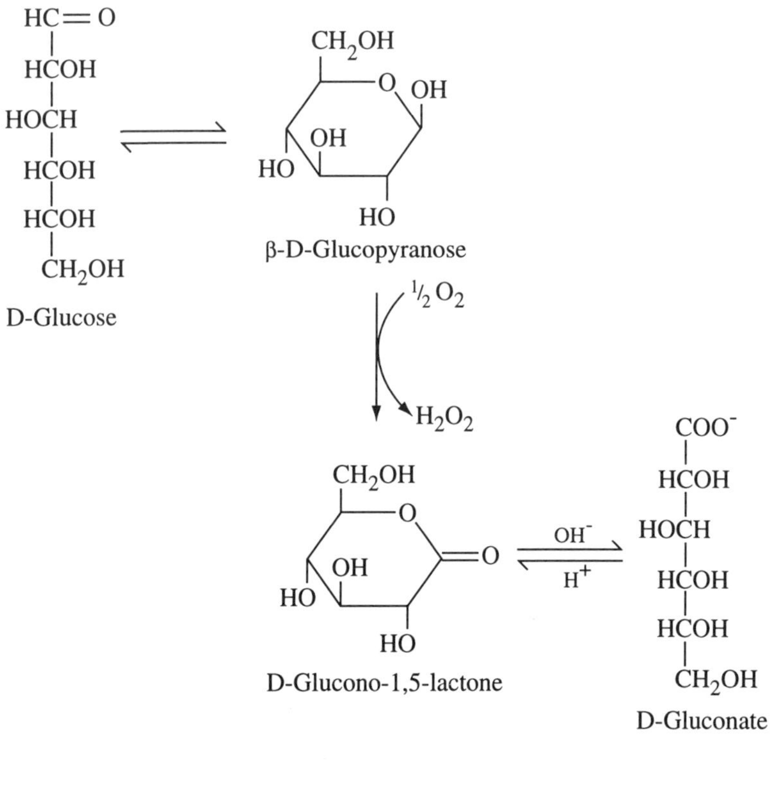

Lactones

Aldonic acids readily undergo intramolecular ester formation to produce a lactone ring. Even vaporization of water from an aqueous solution of an aldonic acid yields the lactone. While the six-membered 1,5-lactone ring can be formed, the five-membered 1,4-lactone ring (γ-lactone ring) is usually more stable and often the only product isolated.

It is obvious that, because of rapid equilibrium between an aldehydo sugar and its pyranose and furanose ring forms and equilibrium between an aldonic acid and its 1,4- and 1,5-lactone ring forms, oxidation of a sugar in any solution form gives the same mixture of products at equilibrium.

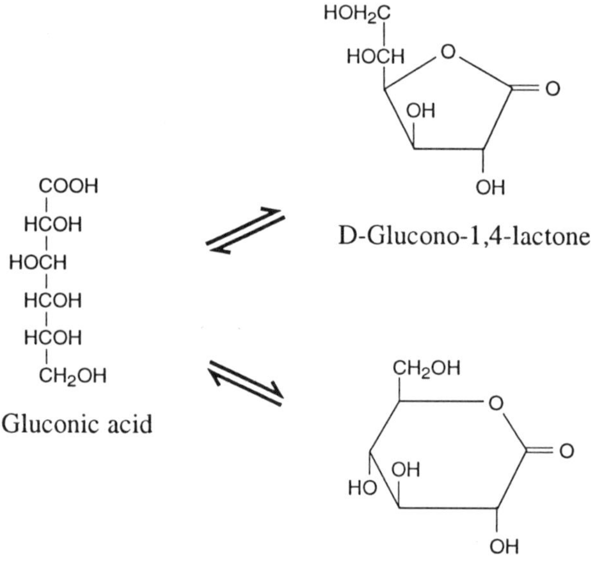

D-Gluconic acid ⇌ D-Glucono-1,4-lactone

⇌ D-Glucono-1,5-lactone

D-Glucono-delta-lactone (GDL), properly D-glucono-1,5-lactone, undergoes hydrolysis to the open-chain carboxylic acid in water in about 3 hr at room temperature, effecting a decrease in pH. During the slow hydrolysis, the initial sweet taste of the solution gradually changes to a slightly acidic taste. The slow acidification from the slow hydrolysis and mild taste makes GDL unique among food acidulants

A primary use of GDL is in the baking industry, where it is utilized as a leavening agent and a preservative. As a leavening agent, it is used to neutralize sodium bicarbonate and release carbon dioxide. Because cold temperatures have a negative effect on normal baker's yeast, GDL is often the leavening agent used in refrigerated and frozen dough products. It also prevents discoloration of refrigerated dough. GDL acts as a preservative by inhibiting microbial growth in bakery fillings. In fish cake and surimi, it lowers the pH, enhancing the action of preservatives such as sorbic acid; inhibits protein degradation; and aids in maintaining color. In meat products, GDL reduces the amount of nitrite required, accelerates the curing process, and lengthens shelf life by inhibiting the growth of lactic acid bacteria. It finds additional applications as an acidulant in improving color stability and firmness in canned and frozen vegetables, in improving the color and texture of pasta and rice, in being a partial replacement for vinegar in salad dressings and sauces, and in being a protein coagulant in the manufacture of cheese and tofu. As already described in the section on glucose oxidase, GDL is formed directly by oxidation of β-D-glucopyranose using glucose oxidase as catalyst.

Reduction of Carbonyl Groups

Sorbitol and Vitamin C

Hydrogenation is the addition of hydrogen to a double bond. When applied to carbohydrates, it most often entails addition of hydrogen to the double bond between the oxygen atom and the carbon atom of the carbonyl group of an aldehyde or ketone. Hydrogenation of D-glucose is easily accomplished with sodium borohydride, with hydrogen gas in the presence of platinum or palladium, or, as done commercially, with hydrogen gas under pressure in the presence of Raney nickel. The product, D-glucitol, commonly called *sorbitol*, is obtained in nearly 100% yield. Sorbitol, properly D-glucitol, where the *-itol* suffix denotes a sugar alcohol, belongs in the general carbohydrate group called *alditols*, also known as *glycitols, polyhydroxyl alcohols*, and *polyols*. Because it is derived from a hexose, it is specifically a hexitol. While it is found widely distributed throughout the plant world, ranging from algae to higher orders, where it is found in fruits and berries, the amounts present are generally small. It was first discovered in the berries of the mountain ash tree (*Sorbus aucuparia*), hence its common name. It is sold both as a syrup and as crystals and is used as a general humectant. It is a noncariogenic sweetener about one-half as sweet as sucrose. It and other polyols are most widely used to replace the physical bulk of sugar.

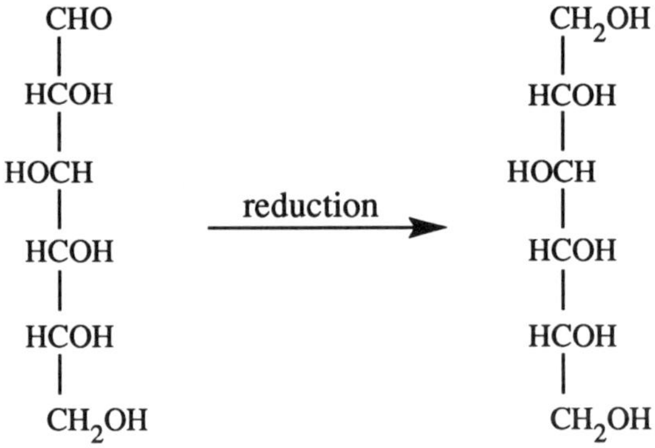

$$\begin{array}{c} \text{CHO} \\ | \\ \text{HCOH} \\ | \\ \text{HOCH} \\ | \\ \text{HCOH} \\ | \\ \text{HCOH} \\ | \\ \text{CH}_2\text{OH} \\ \text{D-Glucose} \end{array} \xrightarrow{\text{reduction}} \begin{array}{c} \text{CH}_2\text{OH} \\ | \\ \text{HCOH} \\ | \\ \text{HOCH} \\ | \\ \text{HCOH} \\ | \\ \text{HCOH} \\ | \\ \text{CH}_2\text{OH} \\ \text{D-Glucitol} \\ \text{(Sorbitol)} \end{array}$$

The largest quantity of sorbitol is used in toothpaste, where it acts as a noncariogenic humectant and plasticizer and imparts a cool sweet taste. The major food use of sorbitol is in sugarless gums, mints, candies, and cough drops. It is also used in nondietetic foods such as shredded coconut, glazed and dried fruits, baked goods, and gelatin products for its humectancy and bodying effect. Its ability to act as a cryoprotectant makes it useful in surimi. Sorbitol is the starting material for preparation of sorbitan esters (see Ethers), which are useful as nonionic food emulsifiers.

Sorbitol is also used in the synthesis of L-ascorbic acid (vitamin C). L-Ascorbic acid is widely distributed in plants and animals. However, humans, other primates, guinea pigs, bats, birds, and fish lack a liver enzyme, L-gulono-γ-lactone oxidase, that is necessary for synthesis of L-ascorbic acid and require an exogeneous source of the vitamin to survive.

Synthesis of ascorbic acid begins with D-sorbitol, which is biologically oxidized at C-5 with *Acetobacter suboxydans* to produce a 95% yield of L-sorbose, a ketohexose. L-Sorbose is converted in three steps (with more than 90% yield at each reaction) into 2-keto-L-gulonic acid, which, on heating to 100°C, lactonizes to ascorbic acid. High yields at each step of the synthesis allow vitamin C to be a low-cost product.

```
  CHO            CH₂OH
  |              |
 HCOH           HCOH              HO    O   OH
  |              |                 \   / \ /
 HOCH    →     HOCH        →       HO    CH₂OH      →    →
  |              |                  |
 HCOH           HCOH                OH
  |              |
 HCOH           HCOH              L - Sorbose
  |              |
  CH₂OH          CH₂OH

 D - Glucose   D - Glucitol
               (Sorbitol)
```

```
                                          CH₂OH
                                           |
      HO    O   OH                        HCOH   O
       \   / \ /                               \   \
        HO                →                           =O
        |  COOH
        OH                                  HO   OH
 "2 - Keto - L - gulonic acid"         L - Ascorbic acid lactone
```

L-Ascorbic acid is required for collagen formation, fatty acid metabolism, good brain function, and drug detoxification; it prevents scurvy and reduces infection and fatigue. In plants, L-ascorbic acid is involved in cellular respiration, growth, and maintenance of carbon balance. It, or an ester of it, is used in foods as an antioxidant.

Natural L-ascorbic acid is prepared in small commercial quantity from rose hips, persimmon, citrus fruit, and other plant sources.

Mannitol

D-Mannitol can be obtained by hydrogenation of D-mannose and can also be produced by fermentation. Commercially, it is obtained along with D-sorbitol from hydrogenolysis of sucrose (Chapter 3). It develops from hydrogenation of the D-fructose component of sucrose or from isomerization of D-glucose, which can be controlled by the alkalinity of the solution undergoing catalytic hydrogenation.

```
   CH₂OH                    CH₂OH              CH₂OH
    |                        |                  |
    C=O                     HCOH               HOCH
    |                        |                  |
   HOCH     reduction       HOCH       +       HOCH
    |      ─────────►        |                  |
   HCOH                     HCOH               HCOH
    |                        |                  |
   HCOH                     HCOH               HCOH
    |                        |                  |
   CH₂OH                    CH₂OH              CH₂OH
```

D-Fructose   D-Glucitol   D-Mannitol

D-Mannitol, unlike D-sorbitol, is not a humectant. Rather it crystallizes easily and is only moderately soluble. It is used as an anticaking agent and for dusting confectionery products. It is 65% as sweet as sucrose and is used in sugar-free chocolates, pressed mints, cough drops, and hard and soft candies. It occurs in nature as manna on flowering ash, locust, olive, and other trees. Brown seaweeds may contain as much as 25% mannitol on a dry weight basis, both free and in combined forms, and it is present in some grasses.

Xylitol

Xylitol is produced from hydrogenation of D-xylose obtained from hemicelluloses. Its crystals have an endothermic heat of solution and give a cool feel when placed in the mouth. It is used in dry hard candies, particularly mint-flavored ones, and in sugarless chewing gum. It is about 70% as sweet as sucrose (Chapter 3). When xylitol is used in place of sucrose, there is a reduction in dental caries because it is not metabolized by the microflora of the mouth to produce plaques. Normally, caries are produced by *Staphylococcus mutans*, which polymerizes the D-glucose portion of sucrose to a polysaccharide called *dextran* that surrounds the colony, allowing it to use the D-fructose portion of the sucrose molecule as an energy source. Since oxygen is restricted beneath the plaque, hypoxic metabolic products, such as lactic acid, accumulate and demineralize tooth enamel to generate caries (tooth decay). Prevention of caries by replacing sucrose with another sweetener is therefore desirable.

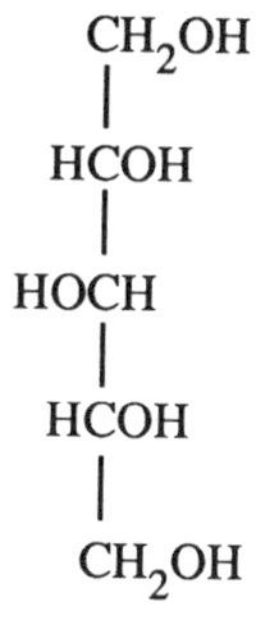

$$
\begin{array}{c}
CH_2OH \\
| \\
HCOH \\
| \\
HOCH \\
| \\
HCOH \\
| \\
CH_2OH
\end{array}
$$

Xylitol

Maltitol

Hydrogenation of maltose, from the partial hydrolysis of corn starch by acid or preferably by the enzyme β-amylase (see Chapters 3 and 6), gives maltitol, a low-cost polyol used in candies and some foods.

See also isomaltitol (Chapter 3).

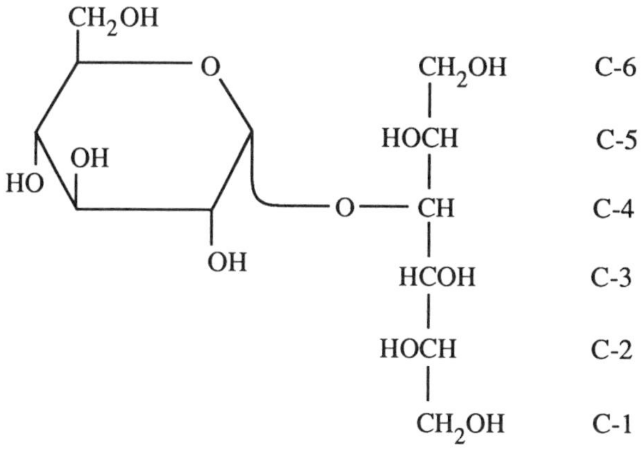

$$
\begin{array}{ll}
CH_2OH & \text{C-6} \\
| & \\
HOCH & \text{C-5} \\
| & \\
CH & \text{C-4} \\
| & \\
HCOH & \text{C-3} \\
| & \\
HOCH & \text{C-2} \\
| & \\
CH_2OH & \text{C-1}
\end{array}
$$

Maltitol

Oxidation of Nonanomeric Hydroxyl Groups

Boiling 30% nitric acid oxidizes both the aldehyde and the primary alcohol groups of an aldose, forming a dicarboxylic acid, a member of the class called *aldaric* (glycaric) *acids*. D-Glucose is converted in 50–65% yield to glucaric acid, commonly called *saccharic acid*. D-Galactose is converted in 75% yield to galactaric acid, commonly called *mucic acid,* a compound sufficiently insoluble that its production has been used to measure the amount of galactose in a product.

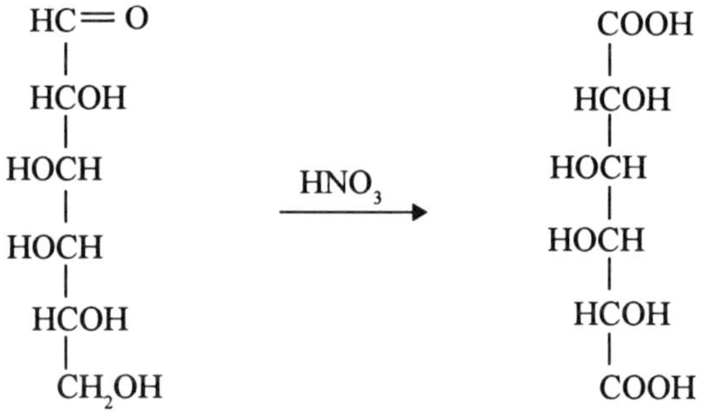

$$\begin{array}{c} \text{HC}=\text{O} \\ | \\ \text{HCOH} \\ | \\ \text{HOCH} \\ | \\ \text{HOCH} \\ | \\ \text{HCOH} \\ | \\ \text{CH}_2\text{OH} \end{array} \xrightarrow{\text{HNO}_3} \begin{array}{c} \text{COOH} \\ | \\ \text{HCOH} \\ | \\ \text{HOCH} \\ | \\ \text{HOCH} \\ | \\ \text{HCOH} \\ | \\ \text{COOH} \end{array}$$

D-Galactose   Galactaric acid
                                (Mucic acid)

Periodate anion (IO_4^-) is a specific oxidant for adjacent hydroxyl groups and is convenient for measuring the number of such groups in a molecule. At each oxidative cleavage of the molecular chain, one molecular portion of periodate is consumed and the secondary alcohol groups involved are converted to aldehyde groups. When three secondary alcohol groups are adjoining, the central one is oxidized twice, resulting in its transformation to formic acid. The most rapid oxidation occurs at pH 3–5. Periodate oxidation is useful in determination of polysaccharide structures.

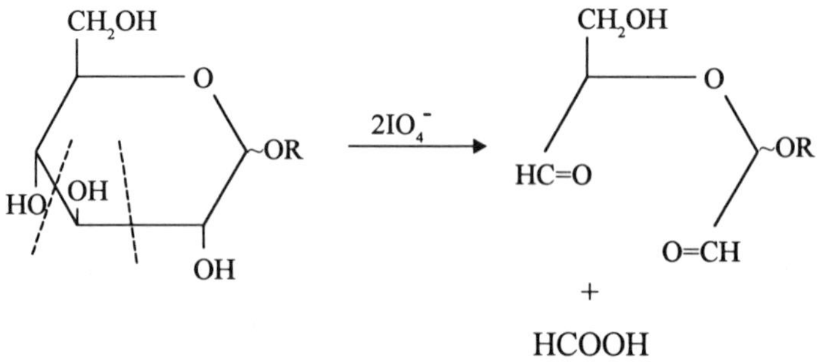

Dinitrogen tetraoxide (N_2O_4) is a fairly selective oxidant for converting primary hydroxyl groups to carboxyl groups. It is most often used experimentally to oxidize polysaccharides, but it is also applicable to glycosides (in which the aldehyde group is protected). The product is a uronic acid derivative. For example, a glycoside of D-glucose is converted into a glycoside of D-glucuronic acid. If a glycuronic acid unit occurs as part of an oligo- or polysaccharide, its glycosidic linkage is rather resistant to hydrolysis, and hydrolysis of the polymer containing it gives a high yield of a dimer, an aldobiouronic acid.[1]

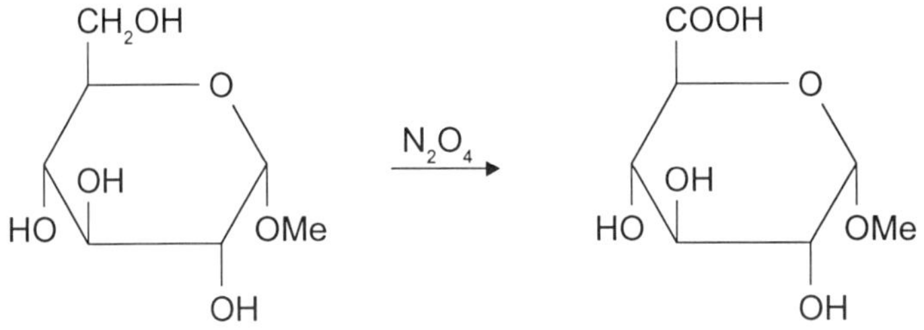

Methyl-α-D-glucopyranoside → (N₂O₄) Methyl-α-D-glucopyranosiduronic acid

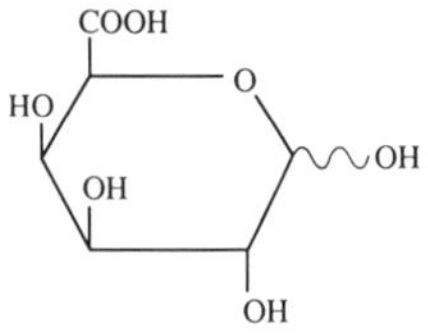

D-Galacturonic acid

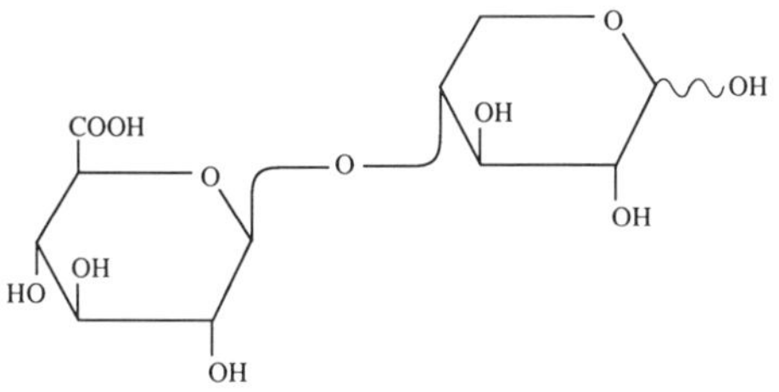

An aldobiouronic acid
(4-O-β-D-Glucuronopyranosyl-D-xylose)

Selective oxidation of primary hydroxyl groups in aqueous solution using oxygen and a platinum or palladium catalyst is used for laboratory preparation of individual uronic acids.

---

[1] An aldobiouronic acid is a disaccharide (Chapter 3) with an uronic acid unit at its nonreducing end.

The enzyme galactose oxidase catalyzes oxidation of the primary hydroxyl group (the C-6 position) of D-galactopyranosides to produce an aldehyde group. Methyl lactoside (the methyl glycoside of the disaccharide lactose [Chapter 3]), for example, can be oxidized to produce an aldehyde group from carbon 6 of the β-D-galactopyranosyl unit. If the aldehyde group is further oxidized, it becomes a carboxyl group in the resulting aldobiouronic acid glycoside. The methyl glycoside group remains easy to hydrolyze, while the D-galactopyranosyl linkage becomes resistant to hydrolysis.

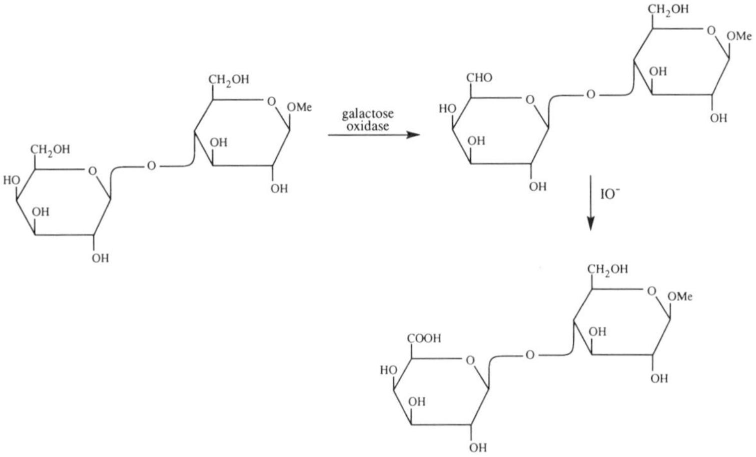

Hydrogen peroxide is a nonspecific oxidant, sometimes used to depolymerize oligo- or polysaccharides. It acts by a free radical mechanism in a reaction catalyzed by ferrous ions, which donate electrons to hydrogen peroxide, resulting in its splitting into hydroxide ions and hydroxyl radicals, as shown in the reaction:

$$Fe^{2+} + HO\text{-}OH \longrightarrow Fe^{3+} + HO\cdot + OH^-$$

The hydroxyl radicals attack the hydroxyl groups of carbohydrates, leading to formation of carbonyl groups.

Esters

The hydroxyl groups of carbohydrates, like the hydroxyl groups of simple alcohols, form esters with organic and some inorganic acids. Reaction of hydroxyl groups with a carboxylic acid anhydride or its chloride (an acyl chloride) in the presence of a suitable base produces an ester:

$$\text{ROH} + \underset{\underset{\text{O}}{\|}}{\text{R'-C}}\text{-O-}\underset{\underset{\text{O}}{\|}}{\text{C-R'}} \text{ or } \underset{\underset{\text{O}}{\|}}{\text{R'-C}}\text{-Cl} \longrightarrow \text{R-O-}\underset{\underset{\text{O}}{\|}}{\text{C-R'}} + \text{HO-}\underset{\underset{\text{O}}{\|}}{\text{C-R'}} \text{ or HCl}$$

For determination of the monosaccharide composition of polysaccharides (Chapter 4) and for analysis of the sugar composition of food products, the monosaccharides that occur naturally or that have been released by acid-catalyzed hydrolysis (Chapter 4) are sometimes first reduced to the corresponding alditols. The alditols are then treated with acetic anhydride in pyridine to produce the fully acetylated derivatives (peracetylated alditols). For example, D-glucose is reduced to D-glucitol, which in turn is acetylated to form D-glucitol 1,2,3,4,5,6-hexaacetate. The alditol percetates are then separated chromatographically, identified, and quantitated. The value in this reaction sequence lies in the fact that each sugar gives a single alditol acetate derivative that is volatile and thermostable and thus can be used for gas-liquid chromatography.

```
    CHO                    CH₂OH                      CH₂OAc
     |                       |                          |
    HCOH                    HCOH                       HCOAc
     |          NaBH₄         |           Ac₂O           |
    HOCH       ------->     HOCH         ------->      AcOCH
     |                       |                          |
    HCOH                    HCOH                       HCOAc
     |                       |                          |
    HCOH                    HCOH                       HCOAc
     |                       |                          |
    CH₂OH                   CH₂OH                     CH₂OAc

  D-Glucose              D-Glucitol            Acetylated D-glucitol
```

Acetates, succinate half-esters, and other carboxylic acid esters of carbohydrates occur in nature. They are found especially as components of polysaccharides (Chapter 4). For example, the polysaccharide xanthan (Chapter 10) contains a 6-O-acetyl-α-D-mannopyranosyl unit. Esters of carboxylic acids are used extensively as hydroxyl protecting and blocking groups in synthetic carbohydrate chemistry because they are easy to prepare, are usually crystalline, and can be removed easily at the necessary time with bases.

Sugar phosphates are common metabolic intermediates. Examples of such compounds are D-glucose 6-phosphate and D-fructose 1,6-bisphosphate.

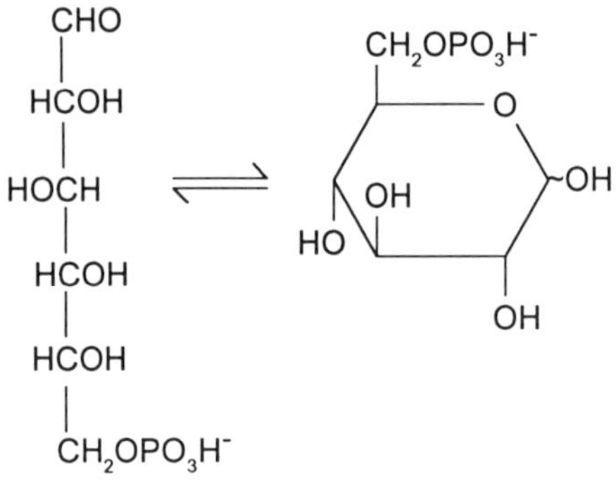
CHO
HCOH
HOCH
HCOH
HCOH
CH2OPO3H-
CH2OPO3H-
O
OH
OH
HO
OH

D-Glucose 6-phosphate

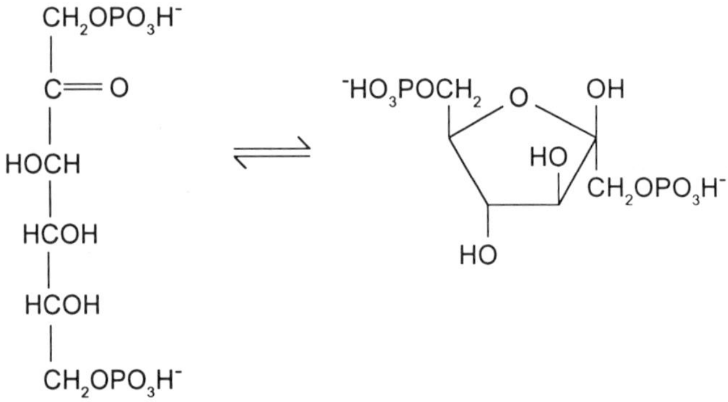
CH2OPO3H-
C=O
HOCH
HCOH
HCOH
CH2OPO3H-
-HO3POCH2
O
OH
HO
CH2OPO3H-
HO

D-Fructose 1,6-bisphosphate

Monoesters of phosphoric acid are also found as constituents of polysaccharides. For example, potato starch contains a small percentage of phosphate ester groups (Chapter 6). Corn starch contains even less, only a trace. Small amounts of phosphate ester groups may also be present in other starches. In the production of modified food starch, corn starch is often derivatized with one or the other or both mono- and distarch ester groups (Chapter 6).

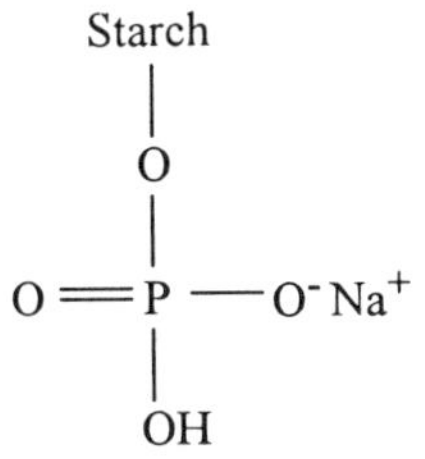

$$\begin{array}{c} \text{Starch} \\ | \\ \text{O} \\ | \\ \text{O}=\text{P}-\text{O}^-\text{Na}^+ \\ | \\ \text{OH} \end{array} \qquad \begin{array}{c} \text{Starch} \\ | \\ \text{O} \\ | \\ \text{O}=\text{P}-\text{O}^-\text{Na}^+ \\ | \\ \text{O} \\ | \\ \text{Starch} \end{array}$$

Monostarch phosphate 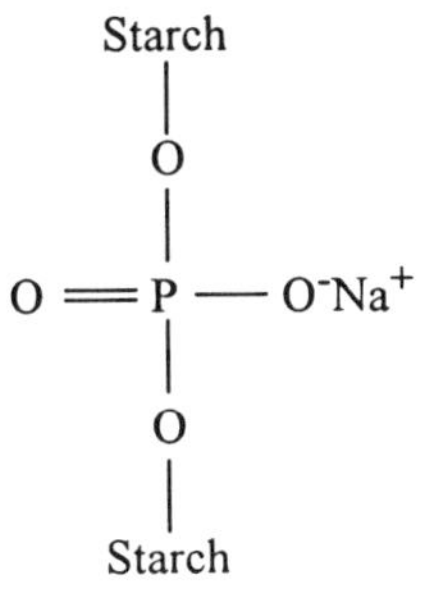 　　　Distarch phosphate
　　　　　　　　　　　　　　(crosslinked starch)

Other esters of starch, most notably the acetate, succinate, and substituted succinate half-esters and distarch adipates, are in the class of modified food starches (Chapter 6). Cellulose esters, especially cellulose triacetate, cellulose diacetate, and cellulose acetate butyrate, are used to make films (Chapter 7). Sucrose (Chapter 3) fatty acid esters are produced commercially as emulsifiers.

The family of red seaweed polysaccharides, which includes the carrageenans (Chapter 11), contains sulfate groups (half-esters of sulfuric acid, $R\text{-}OSO_3^-$).

Ethers

The hydroxyl groups of carbohydrates, like the hydroxyl groups of simple alcohols, can form ethers as well as esters. Ethers of carbohydrates are not as common in nature as are esters. One of only a few examples is 4-*O*-methyl-D-glucuronic acid, a common constituent of hemicelluloses (Chapter 8) and exudate gums (Chapter 14).

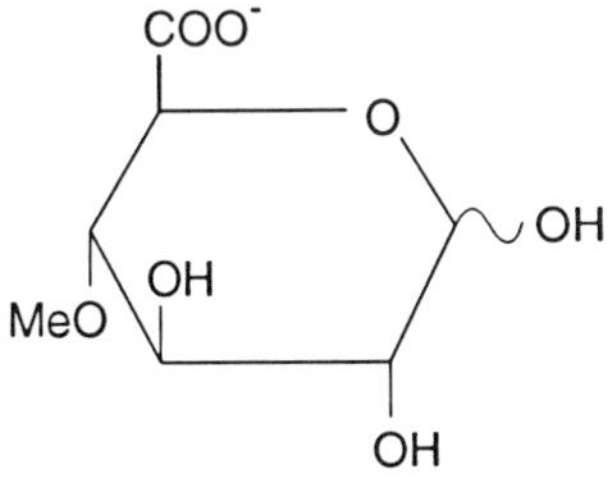

4-*O*-Methyl-D-glucuronate

Polysaccharides are etherified to modify their properties and make them more useful. Examples are the production of methyl (-O-CH_3), sodium carboxymethyl (-O-CH_2-$CO_2^-$$Na^+$), and hydroxypropyl (-O-CH_2-CHOH-CH_3) ethers of cellulose and hydroxypropyl ethers of starch, all of which are approved for food use. (Other ethers of cellulose and starch and ethers of guar gum, along with the ones mentioned, have extensive and widespread use in other industrial applications.)

To determine the chemical structures of polysaccharides, specifically to determine the linkages between sugar (glycosyl) units, that is, whether the linkages are 1→2, 1→3, 1→4, or 1→6, they are subjected to a process called *methylation analysis*. In methylation analysis, all hydroxyl groups not involved in glycosidic bonds are converted to methyl ether groups. The methylated polysaccharide is then subjected to acid-catalyzed hydrolysis (Fig. 2.1). The hydroxyl group involved in the glycosidic linkage can be identified because it is the only one, other than the one involved in ring formation, that is not methylated. This scheme works because methyl ethers, like other ethers, are stable to both acids and bases. The process is demonstrated in Figure 2.1 with methylation analysis of a galactomannan such as guar gum or locust bean gum (see Chapter 9). See Chapter 4 for a more complete description of methylation analysis.

A special type of ether, an internal ether formed between carbon atoms 3 and 6 of a D-galactosyl unit, is found in the red seaweed polysaccharides, specifically agar, furcellaran, κ-carrageenan, and ι-carrageenan (Chapter 11). Such an internal ether is known as a 3,6-anhydro ring (Fig. 2.2), the name of which derives from the fact that the elements of water (HOH) are removed during its formation:

$$R\text{-}OH + HO\text{-}R' \longrightarrow R\text{-}O\text{-}R' + H_2O$$

Several nonionic surfactants based on sorbitol (D-glucitol) are used in foods as water-in-oil emulsifiers and as defoamers. They are produced by esterification of sorbitol with fatty acids. Cyclic dehydration accompanies esterification (primarily at a primary hydroxyl group, that is, C-1 or C-6) so that the carbohydrate (hydrophilic) portion is not only sorbitol, but also its mono- and dianhydrides (cyclic ethers). The products are known as *sorbitan esters*. The product called *sorbitan monostearate* is actually a mixture of partial stearic (C_{18}) and palmitic (C_{16}) acid esters of sorbitol (D-glucitol), 1,5-anhydro-D-glucitol (1,5-sorbitan), 1,4-anhydro-D-glucitol (1,4-sorbitan), both internal (cyclic) ethers, and 1,4:3,6-dianhydro-D-

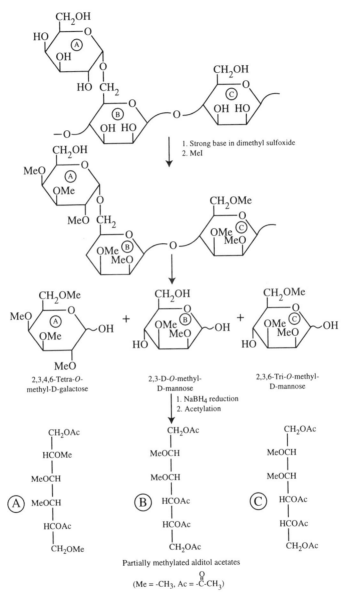

Fig. 2.1. Methylation analysis of a polysaccharide. A is a D-galactopyranosyl unit. B is a 4,6-disubstituted D-mannopyranosyl unit. C is a 4-substituted D-mannopyranosyl unit. The base must be strong enough to bring about ROH → RO⁻.

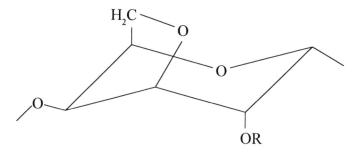

Fig. 2.2. A 3,6-anhydro-α-D-galactopyranosyl unit of κ-carrageenan (R = H) and ι-carrageenan (R = SO$_3^-$).

glucitol (isosorbide), an internal dicyclic ether. Some sorbitan fatty acid esters, such as sorbitan monostearate, sorbitan monolaurate, and sorbitan monooleate, are also modified by reaction with ethylene oxide to produce so-called ethoxylated sorbitan esters, which are also nonionic detergents approved by the U.S. Food and Drug Administration for food use.

Sorbitan Derivatives

Mono-, di-, and tri-esters (Spans)
Polyoxyethylenated derivatives (Polysorbates and Tweens)

Cyclic Acetals

Because they are polyhydroxy compounds, carbohydrates react with aldehydes and ketones to form cyclic acetals (Fig. 2.3). Reactions with reagents such as acetone and benzaldehyde are used in synthetic carbohydrate chemistry as hydroxyl protecting and blocking groups because they react with two hydroxyl groups at once and because they can be removed easily at the necessary time with dilute acid.

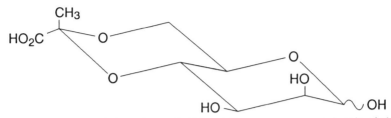

Fig. 2.3. The 4,6-O-(1-carboxyethylidene)-β-D-mannopyranosyl unit of the polysaccharide xanthan. The acetal group (the upper left ring) is a derivative of pyruvic acid, CH_3-CO-COOH.

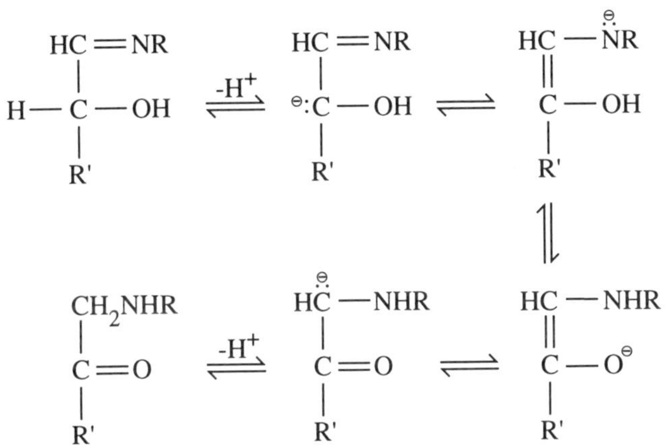

Fig. 2.4. Basic mechanism of the Amadori rearrangement. Compare this with the mechanism for isomerization of aldoses and ketoses (Fig. 1.4).

Browning

The presence of an aldehyde group in sugars gives them great chemical reactivity. Under some conditions, reducing sugars produce brown colors that are important in some foods. Other brown colors obtained upon heating or during long-term storage of foods containing reducing sugars are undesirable.

Maillard Browning

Common browning of foods on heating or on storage is usually due to a chemical reaction between reducing sugars, mainly D-glucose, and a free amino acid or a free amino group of an amino acid that is part of a protein chain. The reaction is named the *Maillard reaction* after the chemist who first examined it in detail. It is also called *nonenzymic (nonenzymatic) browning* to differentiate it from the often-rapid, enzyme-catalyzed browning commonly observed in freshly cut fruits and vegetables, such as apples and potatoes.

When aldoses or ketoses are heated in solution with amines, a variety of reactions ensue, producing numerous compounds (some of which are flavors, aromas, and dark-colored polymeric materials), but both reactants disappear only slowly. The flavors, aromas, and colors, which may be either desirable or undesirable, can be produced by frying, roasting, baking, or storage.

The reducing sugar reacts reversibly with the amine to produce a glycosylamine, as can be illustrated with D-glucose. This undergoes an Amadori rearrangement (Fig. 2.4) to give, in the case of D-glucose, a derivative of 1-amino-1-deoxy-D-fructose. The overall reaction is shown in the following scheme:

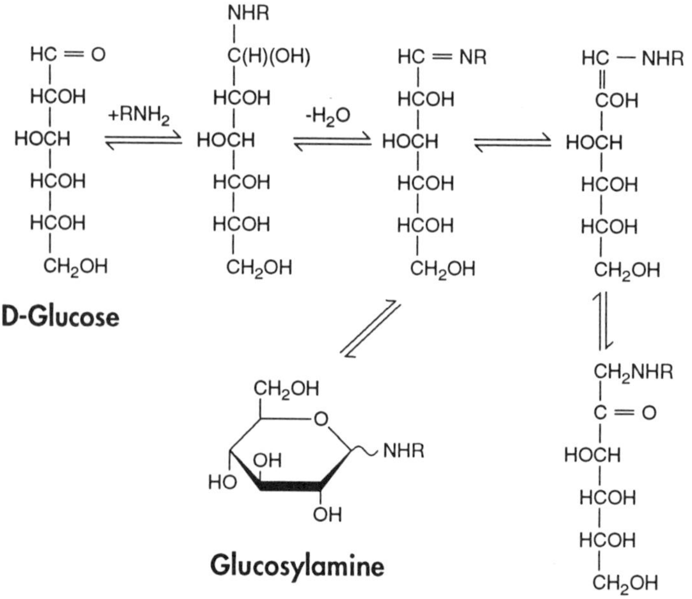

D-Glucose → Glucosylamine → N-Substituted 1-Amino-1-deoxy-D-fructose

Reaction continues, especially at pH 5 or lower, to give a 3-deoxyglycosulose that dehydrates. Eventually a furan derivative forms. For a hexose, this is 5-hydroxymethyl-2-furaldehyde (HMF). Under less acidic conditions (higher than pH 5), the reactive cyclic compounds (HMF and others) polymerize quickly to a dark-colored insoluble material containing nitrogen.

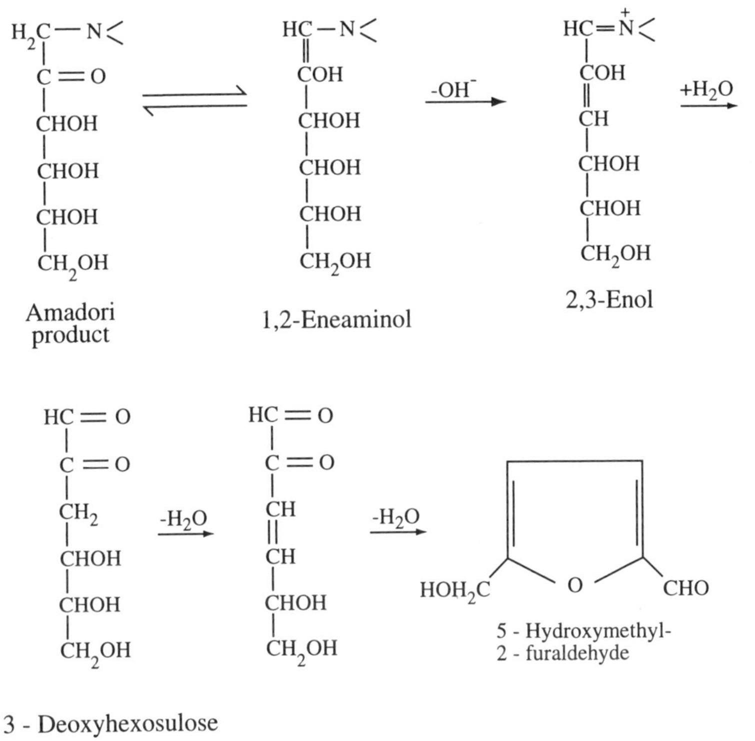

CARBOHYDRATE REACTIONS / 39

$$\begin{array}{c} H_2C-N< \\ | \\ C=O \\ | \\ CHOH \\ | \\ CHOH \\ | \\ CHOH \\ | \\ CH_2OH \end{array} \rightleftharpoons \begin{array}{c} HC-N< \\ \| \\ COH \\ | \\ CHOH \\ | \\ CHOH \\ | \\ CHOH \\ | \\ CH_2OH \end{array} \xrightarrow{-OH^-} \begin{array}{c} HC=\overset{+}{N}< \\ | \\ COH \\ \| \\ CH \\ | \\ CHOH \\ | \\ CHOH \\ | \\ CH_2OH \end{array} \xrightarrow{+H_2O}$$

Amadori product · · · · · · 1,2-Eneaminol · · · · · · · · · · · · 2,3-Enol

$$\begin{array}{c} HC=O \\ | \\ C=O \\ | \\ CH_2 \\ | \\ CHOH \\ | \\ CHOH \\ | \\ CH_2OH \end{array} \xrightarrow{-H_2O} \begin{array}{c} HC=O \\ | \\ C=O \\ | \\ CH \\ \| \\ CH \\ | \\ CHOH \\ | \\ CH_2OH \end{array} \xrightarrow{-H_2O}$$

3 - Deoxyhexosulose

5 - Hydroxymethyl-2 - furaldehyde (HOH$_2$C—furan—CHO)

Maillard browning products, including soluble and insoluble polymers, are found where reducing sugars and amino acids, proteins, and/or other nitrogen-containing compounds are heated together, for example, in soy sauce, nonfat dry milk, and bread crusts. These products are important contributors to the flavor of milk chocolate, which in part is caramel flavor (see Caramel Formation below). The Maillard reaction is also important in the production of caramels, toffees, and fudges, during which reducing sugars also react with milk proteins. Amino acids and furans are often incorporated into the polymers. D-Glucose undergoes the browning reaction faster than does D-fructose. Because the reaction has a relatively high energy of activation, application of heat is generally required for nonenzymic browning. The rate of the Maillard reaction also increases as the water activity (a_w) of a food product increases, reaching a maximum at a_w values in the range of 0.6–0.7. In some cases, further increases in a_w interfere with the Maillard reaction. Thus, for some foods, Maillard browning can be controlled by controlling a_w as well as by controlling reactant concentrations, time, and temperature.

While the Maillard reaction is useful in some cases, it also has a negative side. Reaction of reducing sugars with amino acids destroys the amino acid. This is especially important with L-lysine, an essential amino acid whose ε-amino group can react while the amino acid is a unit of a protein molecule. In addition, mutagenic heterocyclic amines formed by Maillard reactions have been isolated from broiled and fried meat and fish and beef extracts, as well as from model systems.

Caramel Formation

Heating sugars and sugar syrups in the absence of compounds containing amino groups brings about another complex group of reactions termed *caramelization*. Reaction is facilitated by small amounts of acids and certain salts. Mild thermolysis causes anomeric shifts, ring size alterations, and formation and breakage of glycosidic bonds. Mostly, however, thermolysis causes dehydration of the sugar molecule, with the introduction of double bonds or the formation of anhydro rings, as in levoglucosan, a dicyclic acetal. Introduction of double bonds leads to unsaturated rings such as furans. Conjugated double bonds absorb light and produce color. Often, unsaturated rings will condense to polymers yielding useful colors or flavors. Catalysts increase the reaction rate and are often used to direct the reaction to specific types of caramel colors, solubilities, and acidities.

D-glucose $\xrightarrow[-H_2O]{\text{heat}}$ 1,6-Anhydro-β-D-glucopyranose (Levoglucosan)

Brown caramel color made by heating a sucrose (Chapter 3) solution with ammonium bisulfite is used in cola soft drinks, other acidic beverages, baked goods, syrups, candies, pet foods, and dry seasonings. Its solutions are acidic (pH 2–4.5) and contain colloidal particles with negative charges. In this process, the acidic salt catalyzes cleavage of the glycosidic bond of sucrose (see Chapter 3), and the ammonium ion participates in the Amadori rearrangement.

Another caramel color, also made by heating sugar with ammonium salts, is reddish brown, imparts pH values of 4.2–4.8 to water, contains colloidal particles with positive charges, and is used in baked goods, syrups, and puddings. Caramel color made by heating sugar without an ammonium salt is also reddish brown, but it contains colloidal particles with slightly negative charges and has a solution pH of 3–4. It is used in beer and other alcoholic beverages.

The nonenzymic browning caramel pigments are large, polymeric molecules with complex, variable, and unknown structures. It is these polymers that form the colloidal particles. The polymers contain, in addition to ordinary hydroxyl groups, carboxyl, carbonyl, and enolic and phenolic-like hydroxyl groups. Their rate of formation increases with increasing temperature and pH. In the absence of buffering salts, larger amounts of humic substances are formed. Humin has a high molecular weight (its average molecular formula is about $C_{125}H_{188}O_{80}$) and a bitter taste. Therefore, its formation is generally minimized.

Certain pyrolytic reactions produce unsaturated ring systems that have unique flavors and aromas, in addition to substances of color. Maltol (3-hydroxy-2-methylpyran-4-one) and isomaltol (3-hydroxy-2-acetylfuran) contribute to the flavor of bread. 2H-4-Hydroxy-5-methylfuran-3-one has a burnt flavor, as in cooked meat, and can be used to enhance various flavors and sweeteners.

Maltol   Isomaltol   2H-4-Hydroxy-5-methylfuran-3-one

Chapter 3

Oligosaccharides

An oligosaccharide (*oligo-* means *few* in Greek) contains two to 20 sugar units joined by glycosidic bonds. When a molecule contains more than 20 units, it is a polysaccharide (*poly-* means *many* in Greek). In polymer chemistry, the two categories are called *oligomers* and *polymers*, respectively, where *-mer* designates a structure composed of parts (*meros* is the Greek word for *part*).

Disaccharides are glycosides in which the aglycon is a monosaccharide unit. Addition of another monosaccharide unit forms a trisaccharide (a three-unit structure). Progressive addition of glycosyl units to form larger and larger molecules, whether linear or branched, results in tetra-, penta-, hexa-, hepta-, octa-, nona-, and decasaccharides, and so on, containing, respectively, four, five, six, seven, eight, nine, and 10 glycosyl units, and so on.

Oligo- and polysaccharides are composed of glycosyl (saccharide) units joined together in a head-to-tail fashion by glycosidic linkages. As a result, each oligo-and polysaccharide molecule has one reducing (head) and at least one nonreducing (tail) end (Fig. 3.1). Because glycosyl units have multiple hydroxyl groups that can be involved with the anomeric carbon atom of a different glycosyl unit in a glycosidic bond, several units can be joined together, forming branched structures (Fig. 3.1). Oligo- and polysaccharides have one, and only one, reducing end. Branched oligo- and polysaccharides have more than one nonreducing end.

Maltose, obtained by hydrolysis of starch, is an example of a disaccharide. Chemically, the disaccharide maltose is named as a derivative of α-D-glucopyranose (the aglycon [the unit on the right as customarily written]). It is derivatized at the oxygen atom on carbon atom 4, termed O-4, and is therefore 4-*O*-(α-D-glucopyranosyl)-

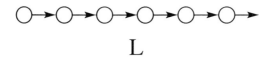

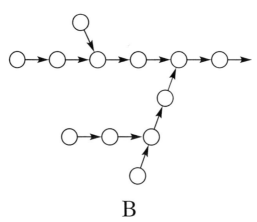

Fig. 3.1. Schematic representations of linear (L) and branched (B) oligosaccharides. Arrows represent the linkages joining the anomeric carbon atom of one glycosyl unit and a hydroxyl group of another. Thus, the end with a free arrow is the end with a free aldehyde (or ketone) group and, therefore, the reducing end. Ends that are glycosyl units without another unit attached to one of their hydroxyl groups by a glycosidic bond (represented by circles only), are nonreducing ends. Thus, the specific branch-on-branch structure presented (B) has one reducing end and four nonreducing ends.

D-glucopyranose. Since the aglycon end has a potentially free aldehyde group, it is the reducing end and will be in equilibrium with α and β six-membered ring forms, as described earlier for monosaccharides. Since O-4 is blocked by attachment of the second D-glucopyranosyl unit, a furanose ring cannot form, and the bulky substituent causes the resulting six-membered ring to be of lower energy and improved stability, thus lowering the number of other possible isomeric forms that might be present. Maltose is a reducing sugar because its aldehyde group is free to react with oxidants, such as Fehling solution, and, in fact, to undergo almost all reactions as though it were present as a free aldose.

Maltose

Because glycosidic bonds are part of acetal structures (Chapter 1), they undergo acid-catalyzed hydrolysis, that is, cleavage in the presence of aqueous acid and heat. However, the reaction is reversible, so sugars can react with each other when in the presence of mild acid and a deficiency of water. As has been seen earlier, sugar acetals or ketals are formed by the reaction of a sugar aldehyde group or a sugar ketone group with two alcohol groups. Since the stable form of sugars contains the hemiacetal ring (involving one of its own hydroxyl groups), only the second reaction is necessary to make a complete acetal or ketal, called a *glycoside*. When the second reacting alcohol group is part of another monosaccharide molecule, the product, a combination of two sugar units, is a disaccharide. Larger structures can also be formed. But this is not the way that either disaccharides or larger oligosaccharides are obtained commercially. Only a few oligosaccharides occur in nature. Most are produced by hydrolysis of polysaccharides into smaller units. However, there is one commercial exception—the production of polymers and oligomers from a monosaccharide by condensation (described below).

The acid-catalyzed combination of two monosaccharide molecules is called *reversion*. Such a reaction joining one monosaccharide molecule with another can occur at any hydroxyl position (for example, at O-2, O-3, O-4, or O-6 of a hexopyranose) and in either anomeric configuration (α or β). Thus, it is theoretically possible to form all possible disaccharides containing only one type of sugar from that sugar alone. For example, a condensation of D-glucose obtained from maltose or starch will produce some disaccharide molecules containing a glycosidic union with a configuration opposite to that of maltose, that is, 4-*O*-(β-D-glucopyranosyl)-D-glucopyranose (cellobiose), which has the glycosidic linkage in the β-D or equatorial position β-Cellobiose

This more or less random reaction between sugar molecules produces a complex mixture of disaccharides, with the amount of each depending on its comparative stability. In addition, the glycoside-forming reaction can continue, especially under water-limiting conditions, to produce trisaccharides, tetrasaccharides, and saccharides of higher members. Such mixtures produced by acid catalysis contain a variety of glycosidic linkages formed in somewhat random fashion. They are rather highly branched because more than one sugar molecule can react with the hydroxyl groups on another sugar unit, which perhaps is already derivatized and has a sugar unit attached to another hydroxyl group. Such branched structures and extensive oligomer formation occur only under nearly dry conditions, because, in the presence of water, hydrolysis of glycosidic bonds also takes place, reversing glycoside formation. One such randomly branched structure is made commercially from D-glucose, sorbitol (Chapter 2), and citric acid (as a catalyst, but it is also a reactant, occurring as an ester group in the final product) and is sold as Polydextrose. The degree of polymerization in Polydextrose is low.

Maltose

Maltose is readily produced by hydrolysis of starch using the enzyme β-amylase, which releases disaccharide (maltose) units sequentially from the nonreducing ends of the starch polymers (Chapter 6). Maltose occurs only rarely in plants, and even then, it results almost entirely from partial hydrolysis of starch. Because of the randomness of acid-catalyzed hydrolysis and consequent low yields of disaccharide, maltose is produced commercially by the specific enzyme-catalyzed hydrolysis of starch, using β-amylase from *Bacillus* bacteria (although the β-amylases from barley seed, soybeans, and sweet potatoes may be used). β-Amylase produces maltose in yields greater than 80%. Higher yields are obtained by use of a mixture of β-amylase and a debranching enzyme (see Chapter 6). Maltose is crystallized easily from aqueous solution as α-maltose monohydrate, even though in solution the ratio of the α to β forms is 1:2. Maltose is used sparingly as a mild sweetener for foods and in pharmaceuticals and as a parenteral injectable for slow release of D-glucose (blood sugar).

Lactose

The disaccharide lactose occurs in milk, mainly free, but to a small extent as a component of higher oligosaccharides. The concentration of lactose in milk varies with the mammalian source from 2.0 to 8.5%. Cow and goat milks contain 4.5–4.8% lactose, human milk about 7%. Lactose is the primary carbohydrate source for developing mammals. In humans, lactose supplies 40% of the energy consumed during nursing. Utilization of lactose for energy must be preceded by hydrolysis to the constituent monosaccharides, D-glucose and D-galactose; hydrolysis is catalyzed by the enzyme lactase, which is present in the small intestine. Milk also contains 0.3–0.6% lactose-containing oligosaccharides, many of which are important as energy sources for growth of a specific variant of *Lactobacillus bifidus*, which, as a result, is the predominant microorganism of the intestinal flora of breast-fed infants.

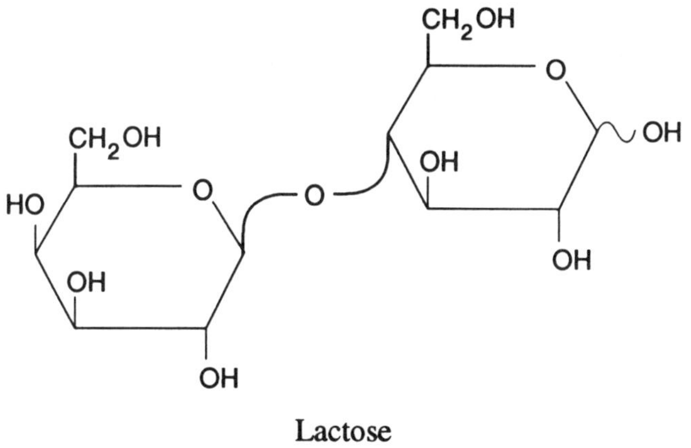

Lactose

Lactose is produced commercially from cow's milk that has had the casein coagulated by heat after adjustment to its isoelectric pH of 4.5–4.7 or by use of rennin, as the first step in the manufacture of cheese. The resulting sweet whey is subjected to ultrafiltration to remove remaining proteins. Then, minerals are removed by ion-exchange, and the lactose solution is concentrated to 50–65% solids to allow it to crystallize or be precipitated. The lactose is redissolved, decolorized with carbon, and recrystallized as β-lactose monohydrate. For every pound of cheese produced, about 9 lb of whey is recovered. Since whey contains about 4.7% lactose, making about 0.4 lb (200 g) potentially available as a by-product from the manufacture of each pound of cheese, and since whey from cheese production amounts to over 23 billion pounds per year in the United States, there is an enormous potential source for lactose, but unfortunately little commercial use at this time.

Lactose is 4-O-(β-D-galactopyranosyl)-D-glucopyranose. Like maltose, it is a reducing sugar with a potentially free aldehyde group. It hydrolyzes in acid solution to produce one molecule of D-galactose and one molecule of D-glucose. Like maltose and other reducing sugars, it exists in water solution in equilibrium with its anomeric forms.

Lactose is used to a small extent in toppings, icings, pie fillings, confections, and ice creams. It contributes body to foods, with only about 40% of the sweetness of sucrose, and finds application because of its enhancement of colors and flavors. It is present in about 20% of tabletted prescription drugs and in about 6% of over-the-counter drugs, for which it provides bulk and rapid dissolution.

Lactose is ingested in milk and other unfermented dairy products, such as ice cream. The fermented dairy products, such as most yogurt and cheese, contain less lactose because, during fermentation, some of the lactose is converted into lactate. Lactose stimulates intestinal adsorption and retention of calcium.

Lactose is not digested until it reaches the small intestine, where the hydrolytic enzyme lactase is located. Lactase (a β-galactosidase) is a membrane-bound enzyme located in the brush border epithelial cells of the small intestine. It catalyzes the hydrolysis of lactose into its constituent monosaccharides, D-glucose and D-galactose:

$$\text{lactose} \xrightarrow{\text{lactase}} \text{D-glucose} + \text{D-galactose}$$

Of the carbohydrates, only monosaccharides are absorbed from the intestines. Both D-glucose and D-galactose are rapidly absorbed and enter the blood stream.

If for some reason the ingested lactose is only partially digested, that is, only partially hydrolyzed, or is not hydrolyzed at all, a clinical syndrome called *lactose intolerance* results. The symptoms of this syndrome are abdominal distention, cramps, flatulence (gas), and diarrhea. This may happen if the person consuming the lactose has a deficiency of lactase, in which case some lactose remains in the lumen of the small intestine. The presence of lactose tends to draw fluid into the lumen by osmosis. It is this fluid that leads to the abdominal distention, cramps, and diarrhea. From the small intestine, the lactose passes into the large intestine (colon), where anaerobic bacteria ferment it to lactic acid (present as the lactate anion) and other short-chain acids. The increase in the concentration of molecules increases the osmolality of the intestinal fluid and results in still greater retention of water. The acidic products of fermentation lower the pH and irritate the lining of the colon, leading to increased movement of the contents. Diarrhea is caused by the retention of fluid and the increased movement of the intestinal contents. The gaseous products of fermentation (carbon dioxide, hydrogen, and methane) cause bloating.

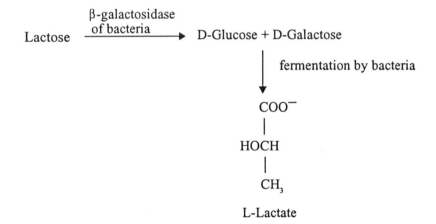

Lactose intolerance is not usually seen in children until after about six years of age. At this age, the percentage of lactose-intolerant individuals begins to rise and increases throughout the life span, with the greatest incidence in the elderly.

The incidence of lactose intolerance varies among ethnic groups. The difference in incidence among whites of western European descent and African Americans is large. By 12 years of age, 45% of

African Americans develop the symptoms of lactose intolerance; among teenagers, the incidence climbs to 70%; and by adulthood, 80% of the African American population shows symptoms of lactose intolerance. Lactose intolerance is also high among Asian Americans (65–100% incidence). Among whites of Western European ancestry, the peak incidence in adulthood is about 6–25%. The incidence among Native Americans is 50–75%; 47–74% of Mexican Americans are estimated to exhibit symptoms. In every case, the variation in estimates is caused by the fact that most cases are unreported. This information indicates that the presence or absence of lactase is under genetic control.

There are two ways to overcome the effects of lactase deficiency. One is to remove the lactose by fermentation; that produces yogurt and buttermilk products. The other, reducing the lactose concentration, is done by adding lactase to milk. However, both products of hydrolysis, D-glucose and D-galactose, are sweeter than lactose, and at about 80% hydrolysis, the taste change becomes too evident. Therefore, in most of these products, the lactose is as close as possible to the 70% government-mandated minimum for reduced-lactose milk. In a novel technology under development, live yogurt cultures are added to refrigerated milk. The bacteria remain dormant in the cold and do not change the flavor of the milk, but, upon reaching the small intestine, they release lactase.

Other carbohydrates that are not completely broken down into monosaccharides by intestinal enzymes are not absorbed and pass into the colon. There they also are metabolized by microorganisms, producing lactate and gas. Again, diarrhea and bloating result. This problem can occur from eating beans, because beans contain a trisaccharide (raffinose) and a tetrasaccharide (stachyose) (see Sucrose below) that are not hydrolyzed to monosaccharides by intestinal enzymes and thus pass into the colon, where they are fermented.

On hydrogenation of lactose with hydrogen and Raney nickel, lactitol, 4-O-(β-D-galactopyranosyl)-D-glucitol, is obtained. This derivative is not absorbed from the human small intestine and is metabolized by microorganisms of the large intestine to produce lactic and acetic acids that, by their hydrophilic, water-binding character, facilitate bowel movement; however, in excess amounts, they produce diarrhea. Lactitol is used in controlled amounts to soften the stool. A product with similar physiological action is made from lactose by isomerizing it in alkali to lactulose, where the D-glucose reducing end is converted to D-fructose.

Sucrose

Sweetness first came to humans in the form of sweet fruits and honey. Many thousands of years went by before another source of sweetness was found in sugar cane, first cultivated on the alluvial soil along the Ganges in what is now the State of Bihar. Its production spread slowly, reaching China about the first century B.C. Early Greeks and Romans went without sugar, but sugar cane spread to Israel, Lebanon, and Syria and was brought to Europe by the Crusaders who, like Pliny the Elder in the first century, called it "a kind of honey made from reeds." In Europe, sucrose, commonly called simply *sugar* or *table sugar*, was a delicacy enjoyed only by the nobility for several centuries, and not until the Spanish and Portuguese brought sugar cane to the Americas did the development of large plantations manned by slave labor bring the price to an affordable level. The per-person daily utilization of sucrose in the United States averages about 160 g; however, some of this sucrose is used in fermentations, in bakery products (where it is largely used up in fermentation), and in pet food. The actual average daily amount consumed by individuals in foods and beverages is about 55 g (20 kg or 43 lb per year).

The structure of sucrose, a disaccharide, provides an exception to the general rule for structures of oligo- and polysaccharides. In

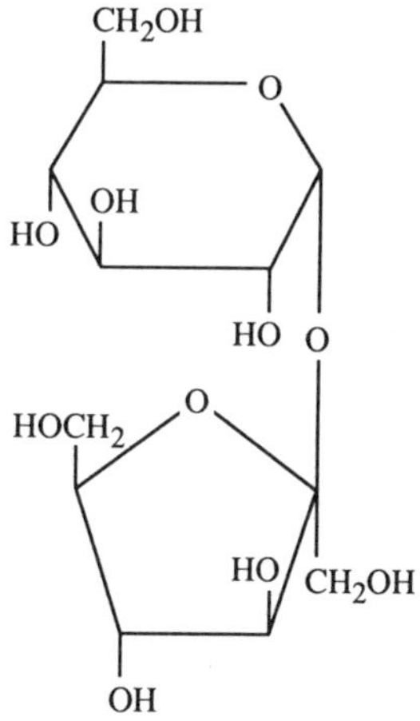

Sucrose

sucrose, the constituent glycosyl units, an α-D-glucopyranosyl unit and a β-D-fructofuranosyl unit, are linked head-to-head, that is, reducing end to reducing end, rather than the very much more common and more general head-to-tail type of linkage. Since both the aldehydo group of the D-glucosyl unit and the keto group of the D-fructosyl unit are covalently bound in a mutual glycosidic bond, sucrose has no reducing end. Therefore, it does not react with Fehling solution or Tollens reagent in normal tests for detection of aldehydes or ketones and is called a nonreducing sugar.

The glycosidic bond between the two sugar rings is of high energy and is unstable, partly due to the strained fructofuranosyl ring. As a result, sucrose is very easily hydrolyzed, even in the presence of very dilute acid, to give an equimolar mixture of D-glucose and D-fructose, termed *invert sugar*.[1] Enzymes also catalyze the hydrolysis of sucrose into D-glucose and D-fructose. The sucrase of the human intestinal tract is one such enzyme, making sucrose one of the oligosaccharides that humans can digest and utilize for energy, the others being the disaccharide lactose and oligosaccharides derived from starch (Chapter 6). Sucrase is an α-D-glucosidase because it leaves the oxygen atom with the D-fructose moiety. Invertases of yeast and bacteria leave the oxygen atom with the D-glucose unit and are therefore β-D-fructofuranosidases. Sucrose is stable under alkaline conditions since it is without a free carbonyl group and acetal and ketals are stable to alkali.

$$\text{sucrose} \xrightarrow{\substack{\text{intestinal sucrase} \\ or \\ \text{bacterial invertase}}} \text{D-glucose} + \text{D-fructose}$$

---

[1] Invert sugar is the equimolar mixture of D-glucose and D-fructose formed by hydrolysis of sucrose. The glycosidic linkage in a sucrose molecule, which joins the anomeric carbon atoms of the two hexosyl units in a head-to-head fashion, is much more prone to acid-catalyzed hydrolysis than is the usual head-to-tail type of linkage. Sucrose is so acid-labile that very dilute acid, for example, very dilute acetic acid (dilute vinegar) at the boiling temperature of a sugar solution, is sufficient to bring about some hydrolysis. The name *invert sugar*, and therefore the name of an enzyme that also effects the hydrolysis, namely *invertase*, derives from the fact that early investigators noticed that the specific optical rotation of sucrose ($[α]_D$ = +66.5°) changed to −33.3°, the $[α]_D$ of the equimolar mixture of the two constituent sugars. As a result of the change from a positive (dextrorotatory) rotation to a negative (levorotatory) rotation, they called the process an inversion and the product *invert sugar*.

The disaccharide sucrose is formed from photosynthetic energy absorbed by the chlorophyll of plants. It derives primarily from uridine diphosphate D-glucose (UDPGlc) and D-fructose 6-phosphate (Fru-6-P), produced early in the photosynthetic pathway. These two components are catalytically combined through the action of sucrose phosphate uridylyl transferase. The remaining phosphate ester group is removed by sucrose phosphatase to yield sucrose:

$$\text{UDPGlc} + \text{Fru-6-P} \longrightarrow \text{sucrose phosphate} \longrightarrow \text{sucrose}$$

Synthesized mainly in leaves, the sugar is transported, along with its caloric energy, to all parts of the plant to supply local energy needs and to provide its carbon atoms, through numerous metabolic routes, for synthesis of all compounds and structures of the plant.

Sources of Sucrose

Cane Sugar

There are two principal sources of commercial sucrose—sugar cane and sugar beets. To obtain sugar from sugar cane, 12- to 18-month-old cane is crushed between rolls. A small spray of water helps remove the juice, which contains about 16% sugar. The juice is made slightly alkaline by addition of lime (calcium hydroxide) to prevent hydrolysis of the acid-labile glycosidic linkage, and the mixture is heated to coagulate proteins, including the hydrolyzing enzyme invertase, and to produce a heavy scum or cake containing a variety of extractives from the juice. The mixture is filtered and concentrated under reduced pressure at a carefully controlled temperature to 50% solids. When crystals of about 300-μm diameter develop, they are removed by centrifugation and given a brief wash. The mother liquor is further concentrated to obtain another crop of crystals. Such concentration and crystallization is conducted until impurities build up to the point where the remaining sucrose will not crystallize. Usually this occurs after two or three crops of crystals are obtained. The final mother liquor, termed *black strap molasses*, is a dark, black, heavy, bitter-flavored syrup, high in ash.

Cane must be brought to the mill and crushed as soon as possible to reduce exposure to microorganisms that would quickly begin to metabolize the sugar. Microorganisms hydrolyze sugar by means of the enzyme invertase; some organisms convert a portion of the sugar to dextran, a soluble polysaccharide that thickens the sugar solution and causes clogging of filters and other mechanical problems in the mill.

Raw sugar, the product in the process described above, is brown and is shipped to a refining mill for purification. There it is dissolved and mixed with lime and phosphate to further precipitate and flocculate impurities. The mixture is clarified and decolorized by centrifugation and by filtration through diatomaceous earth and charcoal. Final crystallization under reduced pressure yields pure, white table sugar.

Beet Sugar

Beet sugar is obtained by countercurrent extraction of sugar beet slices, called *cosettes*, until the liquor contains about 12% sugar. The liquor is agitated with lime for several hours; carbon dioxide is bubbled in to neutralize the alkaline solution with formation of calcium carbonate, and the mixture is filtered. The solution is decolorized with sulfur dioxide. Sucrose is crystallized by concentration of the solution under reduced pressure. After removal of the crystals by centrifugation, the mother liquor is again concentrated, and sugar crystals are collected, with repetition of this sequence until crystals no longer form. Additional quantities of sucrose can be obtained from the mother liquor by diluting it to about 7% sugar, cooling it to 12°C (54°F), and adding lime. A compound of lime and sucrose, termed *tricalcium saccharate*, forms and can be separated by centrifugation. The calcium ions are removed as insoluble calcium carbonate after carbon dioxide is bubbled in.

Also present in sugar beet extract are a trisaccharide, raffinose, which has a single D-galactopyranosyl unit attached to sucrose, and a tetrasaccharide, stachyose, which contains an addition D-galactosyl unit. They have a sweetness of about 0.2% that of sucrose.

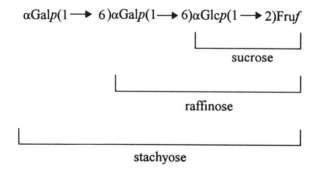

Sucrose is used to make caramel colors and flavors as described in Chapter 2.

Other Sugar Products

Brown Sugar

Commercial brown sugar is made by simple crystallization from lime-treated, but incompletely purified, cane juice or, more often, by treating white sugar crystals with molasses to give them a brown coating of desired thickness. Grades range from light yellow to dark brown. They are high in ash and moisture and are slightly sticky, resulting in their being called *soft sugar* in the industry. Beet sugar molasses is not suitable for brown sugar production.

Powdered Sugar

Confection (powdered) sugar is pulverized sucrose. It usually contains 3% corn starch as an anticaking agent.

Fondant Sugar

To make fondant sugar, which is used in icings and confections, very fine sucrose crystals are surrounded with a saturated solution of invert sugar, corn syrup, or maltodextrin. (See Chapter 6 for the latter two products.)

Transformed Sugar

Agglomerated sucrose crystals are made by causing crystals to stick together with or without the aid of an additive to produce a low-density, rapidly dissolving product. This low-density sugar may be made by extruding crystalline sucrose with sufficient heat and moisture to melt or dissolve crystal surfaces to a very slight degree but sufficiently to promote coherence.

Liquid Sugar

For many food product applications, sucrose is not crystallized; rather, it is shipped as a refined aqueous solution known as *liquid sugar*.

Some Properties of Sucrose Crystals and Solutions

Sucrose and most other low-molecular-weight carbohydrates (for example, monosaccharides, alditols, disaccharides, and other low-molecular-weight oligosaccharides), because of their great hydrophilicity and solubility, can form highly concentrated solutions of high osmolality. Such solutions, as exemplified by pancake and waffle syrups and honey, need no preservatives themselves and can be used not only as sweeteners (although not all such carbohydrate syrups have much sweetness), but also as preservatives and humectants.

A portion of the water in any carbohydrate solution is non-freezable. When the freezable water crystallizes, that is, forms ice, the concentration of solute in the remaining liquid phase increases and the freezing point decreases. There is a consequent increase in viscosity of the remaining solution. Eventually, the liquid phase solidifies as a glass in which the mobility of all molecules becomes greatly restricted and diffusion-dependent reactions become very slow. Water molecules become unfreezable by partial immobilization and/or by an increase in solution concentration to the point that the freezing point is below the freezer temperature. The restricted mobility is a consequence of the high viscosity resulting from concentration (vitrification). In this way, sucrose and other carbohydrates function as cryoprotectants; that is, they protect against the dehydration that destroys structure and texture caused by freezing.

A majority of sucrose molecules in solution have the structure illustrated below, in which the primary hydroxyl group at C-1 of the D-fructofuranosyl unit is strongly hydrogen bonded to the C-2 hydroxyl group of the D-glucopyranosyl moiety to the extent that it is much less reactive than the remaining primary hydroxyl groups. In crystalline sucrose, there is an additional hydrogen bond between the primary hydroxyl group at C-6′ and the ring oxygen atom of the D-glucopyranosyl unit.

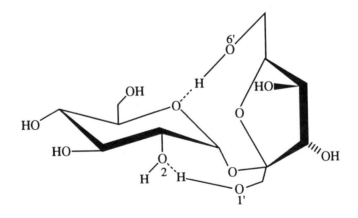

Chocolate contains stable sugar particles in a partially crystalline fat phase. The sugar particles impart non-Newtonian rheology (see Chapter 5) to the chocolate and affect its texture. Particle size distribution is critical. In the United States, the average size of sugar particles is 30–33 μm, with a maximum size of 50 μm. In Europe, the average is 20–23 μm, with a maximum of 40 μm, giving that chocolate a slightly more slippery texture. In both cases, some sugar particles are <1 μm in size, but the human mouth does not detect particles less than 12–15 μm. Small particles require more fat for lubrication and to achieve the required viscosity because of their large surface area. Some crystalline sugar is converted into amorphous sugar during the refining of chocolate. The amorphous sugar absorbs flavor. Also, as described in Chapter 2, the Maillard reaction is important in producing the milk chocolate flavor, and amorphous sugar plays a key role in the development of caramel flavor.

In hard candies, sugar forms glasses in combination with corn syrup. In some confectionary products, sugar forms microcrystals.

Derivatives of Sucrose

Sucrose Esters

Monoesterification of sucrose with a fatty acid, such as stearic acid, produces surfactants. These products are approved as, and used as, food-grade emulsifiers. Fully acetylated sucrose, sucrose octaacetate, is extremely bitter and has been used as a denaturant for ethanol.

Sucrose derivatives that are highly esterified with fatty acids, such as stearic, palmitic and oleic acids, have been developed as low-calorie fat substitutes. Olestra, developed by the Procter & Gamble Company, is such a product. It is not metabolized or absorbed by the body. (A product is metabolizable only when there are four or less ester groups per sucrose molecule.) It provides the same taste and cooking properties as caloric fats and oils and is recommended as a substitute for up to 35% of the fat in home cooking and for up to 75% of the fat in commercial frying of snack food. Olestra is approved for use in specific, limited products.

Sucralose

An interesting commercial derivative of sucrose is a chlorinated product that has a derivatized D-galactopyranosyl unit in place of the normal D-glucopyranosyl unit and that is 600 times sweeter than sucrose. It has a high-quality sweetness and good water solubility and is not hydrolyzed by intestinal invertase. It is hydrolyzed under acidic conditions, but, because of the high electron-withdrawing power of chlorine, it is 60 times more stable to acid than is normal sucrose. Sucralose is recommended for approval and use in a broad array of products, including tabletop sweeteners, beverages, baked goods, chewing gum, dry mixes, processed fruits and spreads, frozen deserts, and salad dressings. It is approved for use in certain specific food products in Canada and some other countries.

Sucralose

Other Oligosaccharides Related to Sucrose

Isomaltulose (Isomalt, Palatinose)

Isomaltulose, 6-O-(α-D-glucopyranosyl)-D-fructose, with half the sweetness of sucrose, can be derived directly from sucrose by enzyme-catalyzed transfer of the D-glucopyranosyl group from O-2 of the D-fructofuranosyl unit to its O-6 position. The enzyme is produced by *Protaminobacter rubrum*. The sweet-flavored, crystalline product is used for making noncariogenic candies, specialty chocolate, chewing gum, and cookies in Europe. Hydrogenation of isomaltulose produces isomaltitol (sold under the trade name Palatinit), an equimolar mixture of disaccharide polyols that is 45% as sweet as sucrose and a substitute for sucrose in making chocolate slabs, drops, and bars, marzipan, and chewing gum.

Leucrose

Leucrose is a reducing disaccharide, 5-O-(α-D-glucopyranosyl)-D-fructose, produced from sucrose by *Leuconostoc mesenteroides*. It has half the sweetness of sucrose.

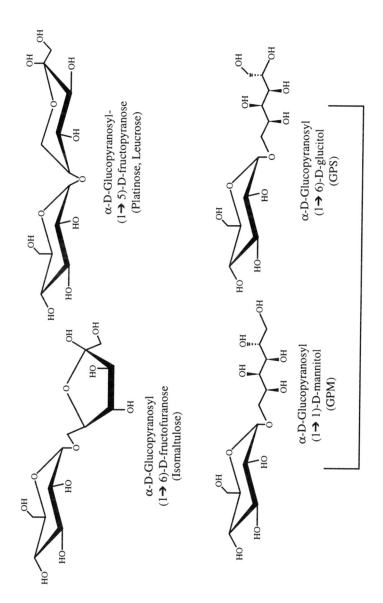

Kestose and Neosugar

The action of invertase, or a specific fungal transferase on a concentrated solution of sucrose, effects cleavage of some of the glycosidic linkages, with transfer of some D-fructofuranosyl units to form a β-D-linkage to O-6 of the D-fructofuranosyl unit of another sucrose molecule. The trisaccharide formed is a *kestose*. Transfer of another D-fructofuranosyl group to the added D-fructofuranosyl unit at the O-

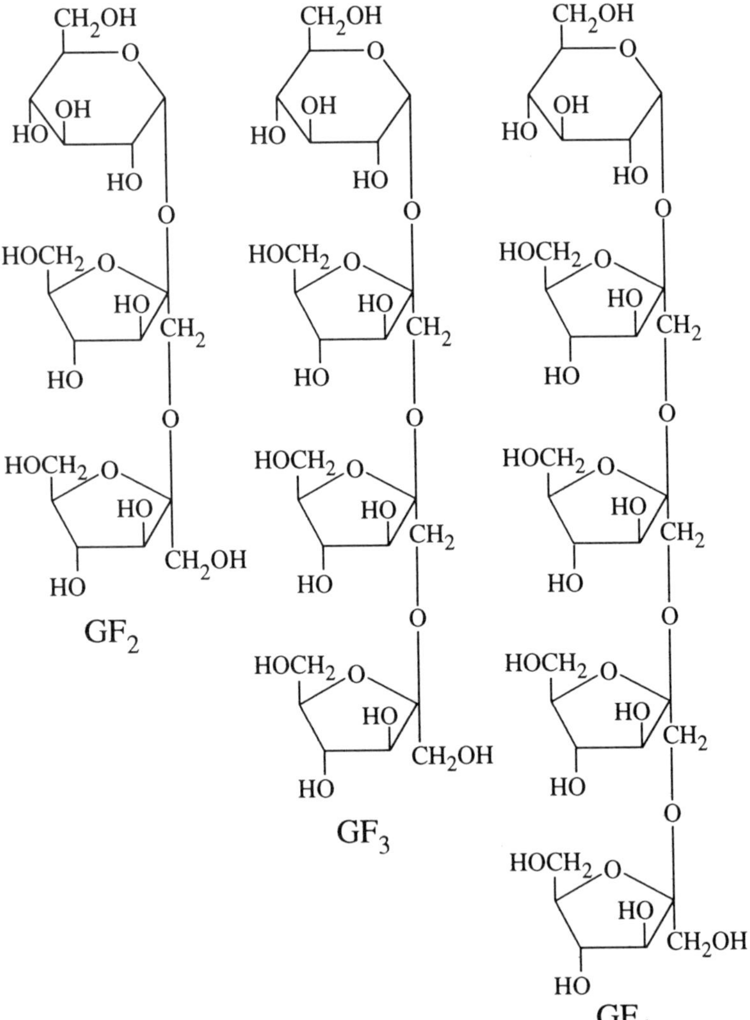

6 position of the kestose extends the chain one unit. This lengthening of the chain produces a mixture of oligosaccharides, called *neosugar*. Neosugar is about half as sweet as sucrose, is said to be noncariogenic and noncaloric, and is used in foods in Japan.

Oligosaccharides Related to Starch

Starches are depolymerized using both acids and enzymes to produce a variety of lower-molecular-weight products, including oligosaccharide products. These products and the processes by which they are made are described in Chapter 6.

Chapter 4

Polysaccharides

Structures

Polysaccharides are polymers of monosaccharides. Thus, they are high-molecular-weight carbohydrate molecules. Like the oligosaccharides, they are composed of glycosyl units in linear or branched arrangements, but most are much larger than the 20-unit limit of oligosaccharides. The number of monosaccharide units in a polysaccharide, termed its *degree of polymerization* (DP), varies with polysaccharide type. Only a few polysaccharides have a DP less than 100; most have DPs in the range of 200–3,000. The larger ones, like cellulose, have a DP of 7,000–15,000. It is estimated that more than 90% of the carbohydrate mass in nature is in the form of polysaccharides (Table 4.1). The general scientific term for polysaccharides is *glycans*, a word derived from *glyc-* for sugar and *-an* for polymer.

If all the glycosyl units are of the same sugar type, they are homogeneous as to monomer units and are called *homoglycans*. Homoglycans can be linear or branched (Table 4.2). Examples of homoglycans are cellulose and starch amylose, which are linear, and amylopectin, which is branched, each of which are composed of D-glucopyranosyl units only.

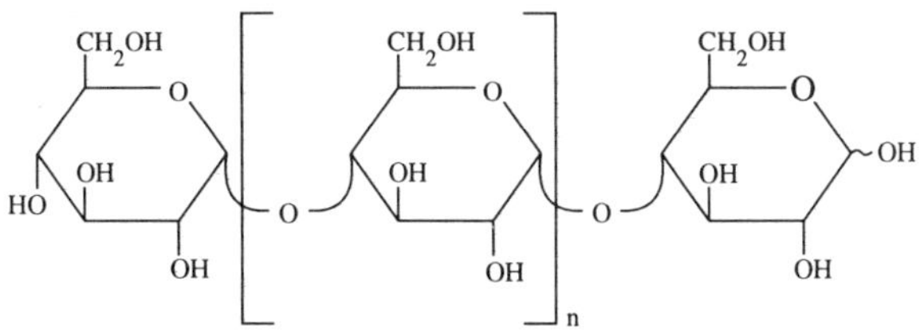

Amylose

TABLE 4.1
Classification of Selected Polysaccharides in Foods by Source

Class	Examples
Algal (seaweed extracts)	Agars, algins, carrageenans, furcellaran
Higher plants	
Insoluble	Cellulose
Extract	Pectins
Seeds	Corn starches, rice starches, wheat starch, guar gum, locust bean gum, psyllium seed gum
Tubers and roots	Potato starch, tapioca starch, konjac mannan
Exudates	Gum arabics, gum karaya, gum tragacanth
Microorganisms (fermentation gums)	Xanthans
Derived	
From cellulose	Carboxymethylcelluloses, hydroxypropylcelluloses, hydroxypropylmethylcelluloses, methylcelluloses
From starch	Starch acetates, starch 1-octenylsuccinates, starch phosphates, starch succinates, hydroxypropylstarches, dextrins
Synthetic	Polydextrose

When a polysaccharide is composed of two or more different monosaccharide units, it is a *heteroglycan*. A polysaccharide that contains two different monosaccharide units is a diheteroglycan; a polysaccharide that contains three different monosaccharide units is a triheteroglycan, and so on. Diheteroglycans generally are either linear polymers of blocks of similar units alternating along the chain, or consist of a linear chain of one type of glycosyl unit with a second type present as single unit branches. Examples of the former type are algins (Chapter 12) and of the latter, guaran and locust bean gum (Chapter 9). (Guaran is the purified polysaccharide from guar gum [Chapter 9].)

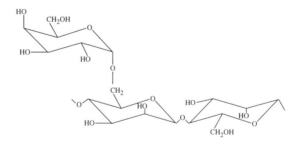

Idealized structure of guaran

TABLE 4.2
Classification of Selected Polysaccharides in Foods by Structure

Classification Schemes	Examples
By shape[a]	
Linear	Algins, amyloses, carrageenans, cellulose, furcellaran, gellan, galactomannans (guar gum, locust bean gum), inulin, pectic acids, pectins
Branched	
Short branches on an essentially linear backbone	Arabinans,[b] arabinogalactans, galactomannans, konjac mannan, protopectins, psyllium seed gum, xanthan, xylans, xyloglucans
Branch-on-branch structures	Amylopectins, arabinoxylans, gum arabics, gum ghatti, gum karaya, gum tragacanth (tragacanthin), okra gum
By monomeric units[c]	
Homoglycans	Amylopectins, amyloses, arabinans, cellulose
Diheteroglycans	Algins, arabinogalactans, carrageenans, furcellarans, galactomannans, glucomannans, inulin, konjac mannan, pectic acids, pectins, xylans
Triheteroglucans	Arabinoxylans, gellan, gum karaya, xanthan
Tetraheteroglycans	Gum arabics, okra gum, psyllium seed gum, xyloglucans
Pentaheteroglucans	Gum ghatti, gum tragacanth (tragacanthin), protopectins
By charge	
Neutral	Amylopectins, amyloses, arabinans, arabinogalactans, cellulose, galactomannans, glucomannans, inulin, konjac mannan, xyloglucans
Anionic (acidic)[d]	Algins, arabinoxylans, carrageenans, furcellarans, gellan, gum arabics, gum ghatti, gum karaya, gum tragacanth (tragacanthin), okra gum, pectic acids, pectins, protopectins, psyllium seed gum, xanthan, xylans

[a] Primary examples. For example, arabinoxylans occur in different architectures, compositions, and charges.
[b] The predominant structure.
[c] Considers only the basic monosaccharide units. A derivatized monosaccharide unit, such as D-galactopyranosyl 6-sulfate unit, is not considered as a unit separate from a D-galactopyranosyl unit, for example.
[d] From the presence of uronic acid, sulfate half-ester, pyruvyl cyclic acetal, or succinate half-ester groups.

Whenever three or more types of monosaccharide units (Fig. 4.1) occur in plant polysaccharides, such as in exudate gums (Chapter 14) and in many hemicelluloses (Chapter 8), the polymers frequently have branch-on-branch structures. Even in such branched structures, simplified arrangements of glycosyl units occur, with one unit type in the main (backbone) chain, which is often a branched, bushlike structure, and other units in short branches. On the other hand, triheteroglycans from bacteria (for example, xanthan [Chapter 10] and gellan) are usually linear or essentially linear molecules. No glycans with more than seven different basic sugar units are known.

While linear glycans are the most abundant in nature because of the enormous quantity of cellulose existing as the main structural element of the cell walls of higher land plants, branched polysaccharides are by far the most numerous overall, occurring in an immense variety of branched forms and with a variety of sugars in their structures.

All polysaccharide preparations contain molecules covering a range of degrees of polymerization. Such preparations are termed *polydisperse*. The molecular weight range may be narrow or broad.

Bacterial polysaccharides, for example, xanthan (Chapter 10) and gellan, and some plant polysaccharides, such as cellulose, are chemically homogeneous. But the chemical structures of most plant polysaccharides vary in linkage types and/or in proportions of monosaccharide constituents from molecule to molecule. The latter are said to be *polymolecular*. Polymolecularity is also introduced when polysaccharides are chemically modified, because the modification takes place at different hydroxyl groups of a glycosyl unit and/or at different locations along chains. Most food gums are both polydisperse and polymolecular. All food gums contain a heterogeneous population of molecules that vary in chemical structure and/or molecular size. This means that, in the case of many gums, a description of the gum consists only of a statistically most probable structure from a population of molecules and that the reported molecular weight of any gum is one of the several types of averages that can be calculated for polymeric molecules.

Polysaccharide preparation, whether in the laboratory for characterization or in commercial production, begins with extraction from the source, in the case of a plant polysaccharide, or isolation from a fermentation culture medium, in the case of a bacterial polysaccharide. In laboratory preparations, extractions from a plant tissue are usually preceded by removal of interfering substances such as lipids

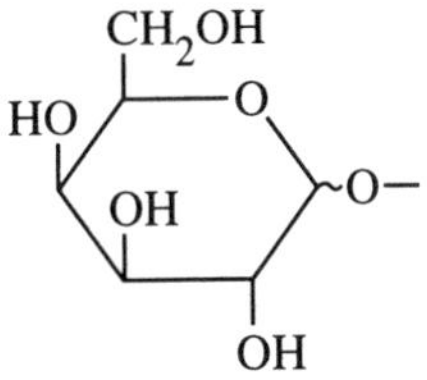

POLYSACCHARIDES / 67

α- and β-D-Galactopyranosyl
(α,βGalp)

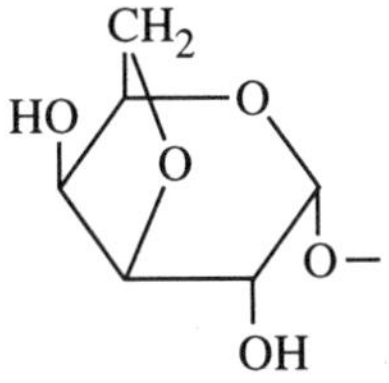

3,6-Anhydro-α-D-galactopyranosyl
(3,6An-αGalp)

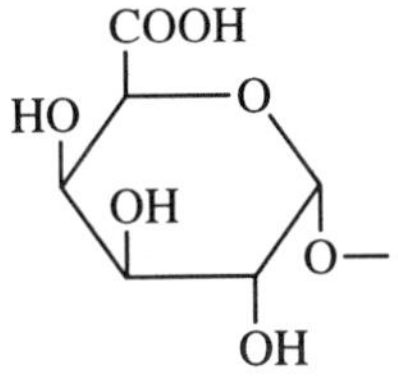

α–D-Galactopyranosyluronic acid
(αGalpA)

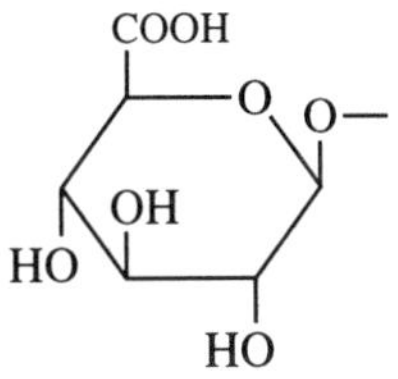

β-D-Glucopyranosyluronic acid
(βGlcpA)

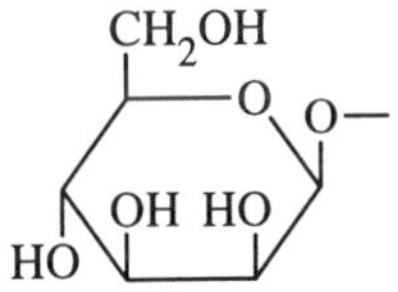

β-D-Mannopyranosyl
(βManp)

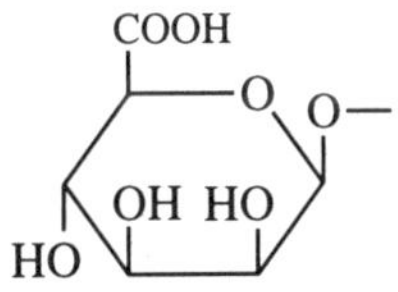

β-D-Mannopyranosyluronic acid
(βManp)

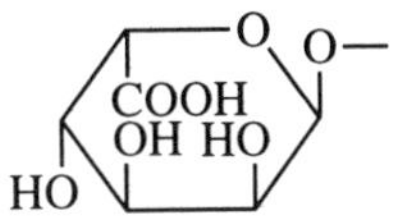

α-L-Gulopyranosyluronic acid
(αLGulpA)

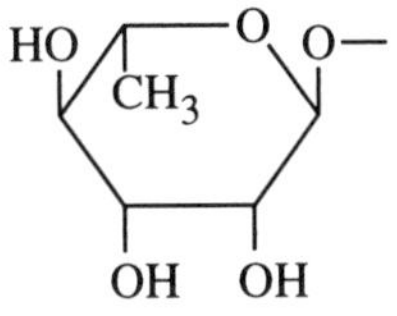

α-L-Rhamnopyranosyl
(αLRhmp)

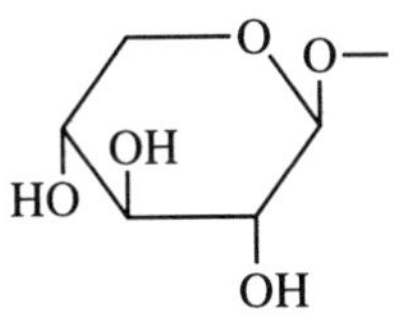

β-D-Xylopyranosyl
(βXylp)

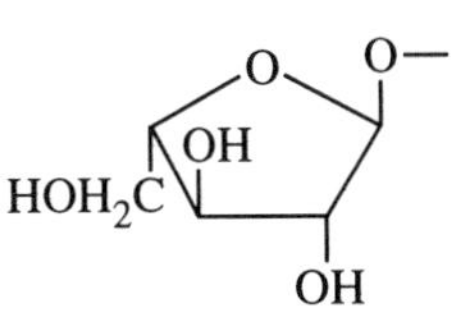

α-L-Arabinofuranosyl
(αLAraf)

Fig. 4.1. Monomer units most often found in polysaccharides.

and lignin. Extraction may be done with water in a few cases but most often involves an alkaline solution. Both extraction and recovery from a fermentation medium are followed by purification and fractionation to separate the desired polysaccharide from noncarbohydrate materials, such as proteins, and from other polysaccharides. Purification most often involves precipitation, sometimes fractional precipitation. Precipitation is usually achieved by addition of a water-soluble alcohol such as ethanol (in a laboratory) or 2-propanol (industrially). It can also be brought about by addition of a complexing agent or by changing the pH of the solution. In the laboratory, chromatographic techniques (size-exclusion and/or ion-exchange) may be used to obtain a reasonably homogeneous preparation.

Structural analysis of a polysaccharide may be undertaken once it is obtained in an acceptable degree of purity. Polysaccharides have a great variety of structures, the only common feature being that each is composed, at least primarily, of monosaccharide units. Structures of polysaccharides can vary with genetics and environmental conditions. Structural characterization involves determination of 1) monosaccharide composition, 2) linkage types, 3) anomeric configurations, 4) presence and location of substituent groups, and 5) degree of polymerization/molecular weight. Because there is such great variability in structures, there is some variability in methods used; however, some generalities can be described.

Structural Characterization

Determination of the monosaccharide composition begins with acid-catalyzed hydrolysis under conditions that give maximum depolymerization and minimum destruction of the sugars. Released monosaccharides are then determined both qualitatively and quantitatively by high-performance liquid chromatography (HPLC) or by gas-liquid chromatography (GLC) after conversion to volatile, thermostable derivatives, often alditol acetates (see Chapter 2).

Linkages are determined by methylation analysis, which can reveal the linkage position, ring size, and the nature of the monosaccharide. Methylation analysis is outlined in Figure 4.2. All exposed hydroxyl groups of the polysaccharide are converted into methyl ethers by treating the polysaccharide, dissolved in dimethyl sulfoxide, with a strong base and reacting it with methyl iodide. Hydrolysis of the completely methylated polysaccharide exposes the hydroxyl groups involved in glycosidic linkages. Of course, the

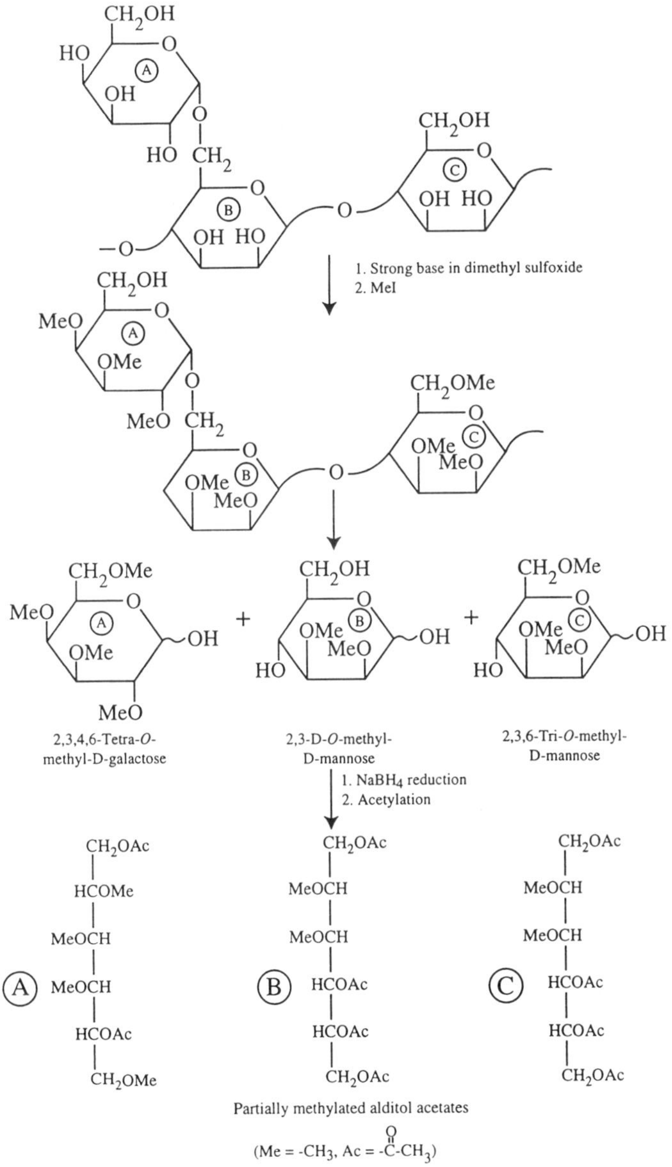

Fig. 4.2. Methylation analysis of an idealized trisaccharide repeating unit of guaran, the polysaccharide of guar gum.

anomeric hydroxyl group of each unit will be involved in a glycosidic bond, so only the other hydroxyl groups are significant. Each of the other hydroxyl groups involved in a glycosidic linkage before hydrolysis will be unmethylated, a characteristic that marks its location. Units that are nonreducing end-units are completely methylated. The released monosaccharides are then reduced to the corresponding alditols, which releases another free hydroxyl group—the one involved in ring formation (the one on C-5 if the unit was originally an aldopyranosyl unit or the one on C-4 if the unit was an aldofuranosyl unit). For separation by GLC, the alditols are acetylated to form partially methylated alditol acetates.

Methylation analysis tells the position of linkages to each monomer unit. It does not tell the sequence. For example, the products produced by methylation analysis of the trisaccharide sequence in Figure 4.2 would also be produced by three other sequences (Fig. 4.3).

In the shorthand notations of oligo- and polysaccharides, the gly-

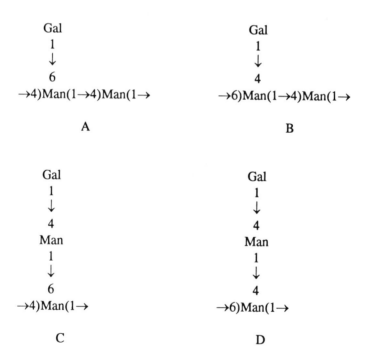

Fig. 4.3. The four structures that will give the three partially methylated alditol acetates shown in Figure 4.2, structure A being the same as the one in Figure 4.2.

cosyl units are designated by the first three letters of their names (except for glucose, which is Glc), with the first letter being capitalized. If the monosaccharide unit is that of a D sugar, the D is omitted; only L sugars are so designated, for example, LAra. The size of the ring is designated by an italicized *p* for pyranosyl or *f* for furanosyl. Uronic acids are designated with a capital A, an L-gulopyranosyluronic acid unit (see Fig. 4.1) being indicated as LGul*p*A, for example. The anomeric configuration is designated with α or β as appropriate. For example, an α-D-glucopyranosyl unit is indicated as αGlc*p*. The position of linkages are designated either as, for example, 1→3 or 1,3, the latter being more commonly used by biochemists and the former more commonly used by carbohydrate chemists. Using this shorthand notation, the structure of lactose is represented as βGal*p*(1→4)Glc or βGal*p*1,4Glc and maltose as αGlc*p*(1→4)Glc or αGlc*p*1,4Glc. (Note that the reducing end is not designated as α or β or as pyranose, except for a specific crystal structure, because, in solution, the reducing-end ring can open and reclose in different ways. Thus, in solutions of oligo- and polysaccharides other than those terminated at the reducing end with a sucrose unit, their reducing end-units occur as mixtures of α- and β-ring forms and the acyclic form, with rapid interconversion between them.)

To determine the monosaccharide sequence, the polysaccharide is partially depolymerized using enzyme- and/or acid-catalyzed hydrolysis to oligosaccharides, the structures of which are then determined. A polysaccharide with a repeating-unit structure such as that in Figure 4.2, which is an idealized structure of guaran, will produce the following trisaccharide fragments:

$$\text{Gal}(1{\rightarrow}6)\text{Man}(1{\rightarrow}4)\text{Man}$$

$$\text{Man}(1{\rightarrow}4)\text{Man}(1{\rightarrow}4)\text{Man}$$

$$\begin{array}{c}\text{Gal}\\1\\\downarrow\\6\\\text{Man}(1{\rightarrow}4)\text{Man}\end{array}$$

Structure A is obviously the only one that could give rise to these products.

The anomeric configuration of the glycosidic linkage also needs to be determined. This can be done on either the intact polysaccharide or on fragments from it. Most commonly, the anomeric configuration is determined using nuclear magnetic resonance or enzymes that are specific for a particular kind of linkage, for example, a β-galactosidase, which is an enzyme that catalyzes only the hydrolysis of a β-D-galactopyranosyl unit at a nonreducing end. After the anomeric configuration is determined, the structure becomes the following:

$$\alpha Gal$$
$$1$$
$$\downarrow$$
$$6$$
$$\rightarrow 4)\beta Man(1\rightarrow 4)\beta Man(1\rightarrow$$

For complete characterization of the polysaccharide, information is needed on its molecular weight and the type and location of any substituent groups. Typical noncarbohydrate substituent groups are acetate ester groups (see xanthan, Chapter 10), sulfate half-ester groups (see carrageenans, Chapter 11), phosphate ester groups (see potato starch, Chapter 6), and pyruvyl cyclic acetal groups (see xanthan, Chapter 10).

Molecular weight is determined by physicochemical methods, including size-exclusion chromatography.

Water Absorption

Polysaccharides modify and control the mobility of water in food systems, so water plays an important role in influencing the physical and functional properties of polysaccharides. Together, polysaccharides and water control many functional properties of foods, including texture.

Most polysaccharides have glycosyl units that, on average, possess three hydroxyl groups. Thus, polysaccharide molecules are polyols in which each hydroxyl group can hydrogen-bond to one or more water molecules. Also, the ring oxygen atom and the glycosidic oxygen atom connecting one sugar ring to another can hydrogen-bond with water. Since each sugar unit in the chain has the capacity to hold water molecules, glycans have a strong affinity for water and readily hydrate with available water. Glycans dried to near zero moisture have a water uptake capacity about equal to that of phosphorus pentaoxide, a strong chemical drying agent. Under normal humidity, glycans generally equilibrate to contain 8–12% moisture. In high-moisture or fluid-water systems, glycan particles take up water, swell, and may dissolve. When dried from highly moist conditions, glycans lose water slowly, allowing hydrated chains time to change from hydrogen bonding with water to hydrogen bonding with hydroxyl groups of other glycan molecules, creating tight intermolecular bonding that produces hard, gritty particles described as hornlike. Such intermolecular associations are difficult to break, even when the particles are placed in excess water. Consequently, such bonding is to be avoided in polysaccharide production and handling.

Water of hydration that is naturally hydrogen-bonded to and thus solvates polysaccharide molecules does not freeze. This water has also been referred to as *plasticizing water*. The motions of these water molecules that solvate carbohydrate molecules are retarded, but they are still able to exchange freely and rapidly with bulk water molecules. Bound water of hydration makes up only a small part of the total water in gels and fresh tissue foods. Water in excess of that involved in hydration is entrapped in capillaries and cavities of various sizes in the gel or tissue.

Solubility and Solution Characteristics

Most, if not all, homoglycans with an essentially linear structure exist in some sort of helical shape, as do most, if not all, polysaccharides, except those with very bushlike, branch-on-branch structures. Certain linear homoglycans, like cellulose (Chapter 7), have a uniform, flat, ribbonlike structure that allows hydrogen bonding to form crystallites.

Cellulose, a linear homoglycan consisting of a chain of β-D-glucopyranosyl units (Chapter 7); mannan, a linear homoglycan of β-D-mannopyranosyl units; and chitin, a linear homoglycan of 2-acetamido-2-deoxy-β-D-glucopyranosyl units, are present in the organisms that make them as highly ordered, highly insoluble products. They contain crystallites intermingled with amorphous regions. In crystallites, the linear, homogeneous chains are packed in crystalline domains of variable lengths and then wander away into less ordered amorphousness. Then, perhaps, they may again enter into and become part of another group of crystallized chain segments. Crystalline domains of this type have no clear surfaces, as seen in familiar simple crystals, and are called *fringed micelles*. It is these crystallites of linear chains that give trunks and branches of trees and shrubs, crab and lobster shells, and insect exoskeletons their great strength, insolubility, and resistance to breakdown, since the crystalline regions are nearly inaccessible to enzyme penetration. However, these highly ordered polysaccharides with orientation and crystallinity are the exception, rather than the rule. Most polysaccharides are not sufficiently crystalline to impart water insolubility, but readily hydrate and dissolve in water, even though they may have ordered conformations that facilitate cooperative interactions between molecular segments (see below and Chapter 5).

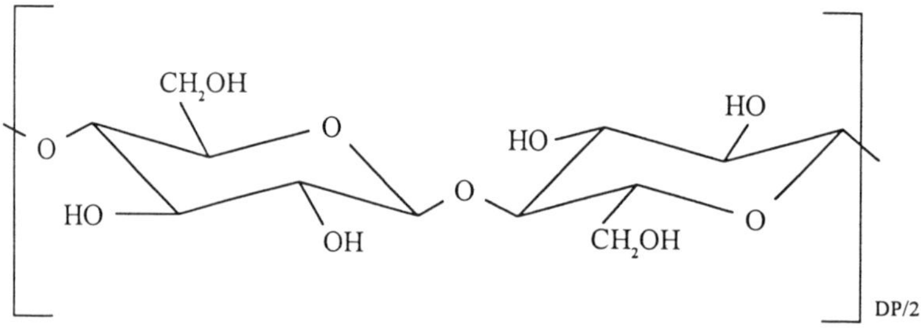

Cellulose

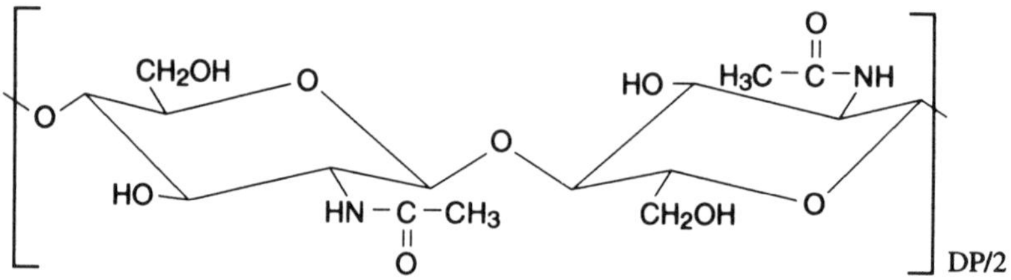

Chitin

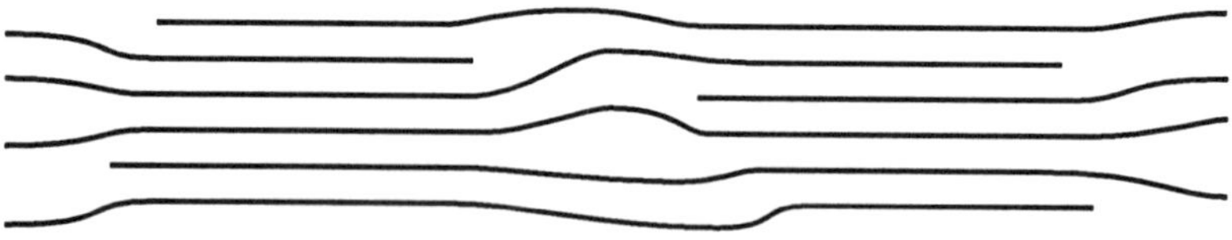

Fringed micelles

Linear diheteroglycans containing nonuniform blocks of glycosyl units and branched glycans cannot form crystalline micelles because chain segments cannot be packed over the lengths necessary to form strong intermolecular hydrogen bonding. Such chains have a degree of solubility. Polysaccharides become more soluble in proportion to the degree of chain irregularity.

Methods for Dissolving Gums

Water-soluble polysaccharides and modified polysaccharides used in food and other industrial applications are known as *gums* or *hydrocolloids*. Commercial gums are powders of various particle sizes, and some, such as carrageenans (Chapter 11) and pectin (Chapter 13), are often sold as mixtures of the gum and finely ground sucrose to facilitate solution preparation and to standardize them with respect to viscosity or gel strength. For homogeneous dispersion, polysaccharides must be added to water under controlled conditions.

If a gum powder is sprinkled onto or poured into slightly stirred water, some will dissolve, but many particles will hydrate quickly on their surfaces to produce a gelatinous coating. This coating slows the rate of water diffusion through the surface layer, leaving a dry interior. Such gel-coated surfaces are sticky, causing the particles to stick together to form clumps called *fish eyes* that may be large or only barely visible.

To prevent this problem and bring about good dissolution of a gum, one of several different procedures may be employed. Finely powdered gum may be sifted slowly into the vortex of rapidly stirred water. Here, high shear forces ensure that particles become dispersed before significant hydration occurs. The shear forces also tear away partially hydrated molecules and prevent the formation of a microgelatinous layer. A high-speed propeller mixer can achieve these conditions if the gum powder is added slowly. Various types of commercial equipment are especially designed for this purpose. One is a powder funnel that allows a fine stream of powder to fall into a jet of turbulent water, giving very rapid mixing. High-shear mixing equipment giving rapid dispersion and mixing of particles in water is essential to making a uniform gum solution.

Alternatively, the gum may be mixed thoroughly with another ingredient such as sugar. This method enhances the probability that gum particles become dispersed before significant hydration occurs. In addition to sugar, other solid or liquid ingredients such as glycerol, propylene glycol, or vegetable oil may be used. However, gums must always be dispersed and dissolved in water before substantial amounts of ingredients that compete for water molecules are added.

Thus, salts, sugars, and other strongly hydrophilic components that hydrate rapidly and compete for water molecules inhibit hydration and formation of a molecular dispersion of gums.

Specially prepared slowly hydrating forms of gums and agglomerated gums are also employed for easier dispersibility.

Even under the best conditions, absolute molecular separation of gum molecules is not readily attained, and some continued hydration of the gum with consequent development of higher viscosity may occur over a period of several minutes to several hours. Raising the temperature after the dispersion is made usually aids in dissolution of the gum.

Properties of Gum Solutions

Polysaccharides (gums, hydrocolloids) are used primarily to thicken and/or gel aqueous solutions and otherwise to modify and/or control the flow properties and textures of liquid food and beverage products and the deformation properties of semisolid foods (See also Chapter 5). They are generally used in food products at concentrations of 0.25–0.50%, indicating their great ability to produce viscosity and to form gels.

The viscosity of a polymer solution is a function of the size and shape (linear or bushlike, for example) of its molecules and the conformations they adopt in the solvent. In foods and beverages, the solvent is an aqueous solution of other solutes.

In solution, molecules oscillate at a minimum energy state because of collisions and thermal energy. The shapes of polysaccharide molecules in solution are a function of oscillations around the bonds of the glycosidic linkages. The greater the internal freedom at each glycosidic linkage, the greater the number of conformations available to each individual segment and the less likely it will be for the chain to adopt a particular shape. Chain flexibility induces the chain to approach disordered or random coil states in solution. However, most polysaccharides form somewhat stiff coils that may be compact or expanded.

Linear polymer molecules in solution gyrate and flex, sweeping out a large space. They frequently collide with each other, consuming energy, creating friction, and thereby producing viscosity. Linear polysaccharides produce highly viscous solutions, even at low concentrations. A 1% water solution of a polysaccharide can easily have a viscosity of 10,000 mPa·sec (centipoise). Viscosity depends on both the DP (molecular weight) and the molecular extension and rigidity, that is, the shape of the solvated polymer chain. Rigidity is influenced by location of the glycosidic bonds and their anomeric configuration, that is, whether the linkages are α or β. Chain extension is dependent, also, on the extent of chain hydration and the amount and type of derivatization of the monomer units.

A highly branched polysaccharide of the same molecular weight as a linear polysaccharide will sweep out far less space. Thus, highly branched molecules collide less frequently and produce a much lower viscosity than linear molecules of the same DP (at the same concentration). This also implies that a highly branched polysaccharide must be significantly larger than a linear polysaccharide to produce the same viscosity at the same concentration (Fig. 4.4).

The ability of polysaccharides to produce high viscosity at low concentrations is a major property of polysaccharides that gives them valuable and widespread industrial use (see also Chapter 5). The relation of chain length to viscosity development was explained earlier (see also Chapter 5). A single cleavage of a glycosidic bond at the center of the chain by hydrolysis, oxidation, or mechanical action, such as shear, produces two polymer molecules of one-half the original molecular weight and much lower viscosity-producing potential. Each half of the molecule, because the radius of the volume it sweeps out upon gyration is one-half that of the original molecule, sweeps out a sphere of but one-eighth the volume of the original molecule, resulting in a viscosity of only one-fourth the

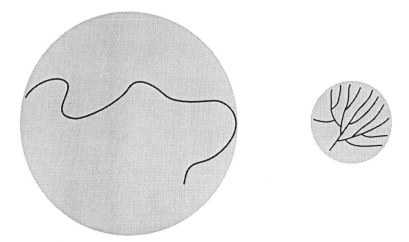

Fig. 4.4. Relative volumes occupied by a linear polysaccharide and a highly branched polysaccharide of the same molecular weight.

original value.[1] As a consequence, care must be taken to prevent chain cleavage during preparation or use of polysaccharides if their ability to produce high viscosity at low concentration is to be maintained. Most gums are available in a wide range of viscosity grades produced by deliberate depolymerization (Chapter 5).

As stated earlier, the degree of chain extension (and coiling) is influenced in part by the configuration of the glycosidic bond. Cellulose, because of its equatorial-equatorial glycosidic bonds, has

---

[1] $r = 1/2$. $r^3 = 1/8$. Therefore, each half would contribute one-eighth the original viscosity, and together they would contribute one-fourth the original viscosity.

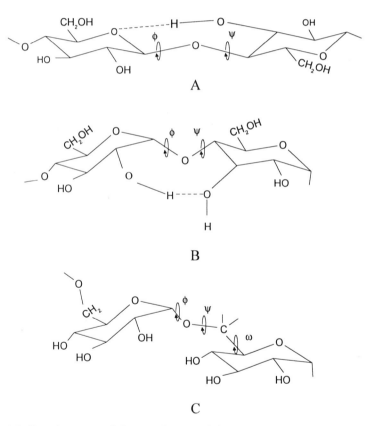

Fig. 4.5. Rotation around the two bonds of the glycosidic linkage (ϕ, ψ) of cellulose (A), amylose (B), and a (1→6)-linked α-D-glucan (dextran, C) and the C-5—C-6 bond (ω) of the latter. Hydrogen bonding stabilizing the conformations of cellulose and amylose.

its (1→4)-linked β-D-glucopyranosyl units positioned so that its adjacent rings can form hydrogen bonds between the ring oxygen atom of one glycosyl unit and the hydrogen atom of the C-3 hydroxyl group of the preceding ring (Fig. 4.5A). These hydrogen bonds hinder free rotation of the rings on their connecting glycosidic bonds and result in stiffening of the chain. The flat, ribbonlike character of the entire molecule allows adjacent cellulose chains to fit closely together in ordered crystalline regions.

In comparison, a chain of (1→4)-linked α-D-glucopyranosyl units (as in the amylose of starch, Fig. 4.6) does not form a ribbonlike structure. Rather the axial-equatorial linkages cause the molecules to coil (see Figure 6.3). The coiled conformation is stabilized by hydrogen bonding between the hydroxyl group on C-2 of one unit and that on C-3 of the preceding unit (Fig. 4.5B)

Nonuniformity along the chains decreases intermolecular interactions and allows greater chain folding. A solvated stringlike molecule without strong intra- and intermolecular hydrogen bonding naturally forms a random structure of, on average, a football-like shape. A coiled configuration is easily adopted by (1→6)-linked polysaccharides because the extra bond between sugar rings moves them too far from each other to form hydrogen bonds between successive units that could restrict ring rotation; this provides an extra degree of flexibility (Fig. 4.5C). Derivatization of chain units (either naturally or industrially) and chain branching hinder association of chain segments and increase solution stability.

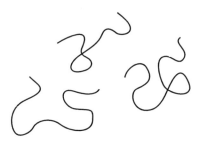

Linear polymer chains bearing only one type of ionic charge (almost always a negative charge, derived, for example, from ionization of carboxyl or sulfate half ester groups in the case of polysaccharides) assume an extended configuration due to repulsion of like charges. Therefore, these polymers tend to impart high viscosity to their solutions.

Polysaccharides do not significantly increase the osmotic pressure or depress the freezing point of water because they are large, high-molecular-weight molecules used at low concentrations. When a starch solution is frozen, a two-phase system of crystalline water (ice) and a glass consisting of about 70% starch molecules and 30% nonfreezable water is formed. As in the case of solutions of low-molecular-weight carbohydrates, the nonfreezable water is in a highly concentrated polysaccharide solution in which the mobility of the water molecules is restricted by the extremely high viscosity. However, while most polysaccharides provide cryostabilization by producing this matrix that limits molecular mobility, there is evidence that some provide cryostabilization by restricting ice crystal growth through adsorption to active growth sites on the crystal surface. Some polysaccharides may be ice nucleators.

Both high- and low-molecular-weight carbohydrates (Chapter 3) are generally effective in protecting food products stored at freezer temperatures (typically –18°C) from destructive changes in texture and structure. In both cases, the improvement in product quality and storage stability results from controlling both the amount and the properties of the freeze-concentrated, amorphous matrix around ice crystals.

Molecular Associations

Many linear glycans, when dissolved in water by heating or by the aid of solubilizing agents, such as a base (which is then neutralized), form unstable molecular dispersions that rapidly precipitate or gel. This occurs as segments of the long molecules, which are undergoing Brownian movement, collide and form intermolecular bonds over the distance of a few units. Initial short alignments then extend in a zipperlike fashion to greatly strengthen intermolecular associations. Segments of other chains colliding with

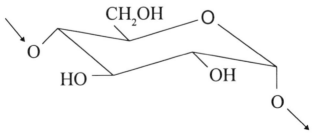

Fig. 4.6. A rough conformational depiction of an amylose molecule.

this organized nucleus bind to it, increasing the size of the ordered, crystalline phase. Linear molecules continue to bind, producing a fringed micelle that may reach a size at which gravitational forces cause precipitation. For example, starch amylose, when dissolved in water with the aid of heat and then cooled to below 65°C, undergoes molecular aggregation and precipitates, a process called *retrogradation*. During cooling of bread and other baked products, amylose molecules associate to produce firming. Over a longer storage time, the branches of amylopectin associate to produce staling (see starch retrogradation in Chapter 6).

In general, all linear, neutral homoglycans have an inherent tendency to associate and partially crystallize. However, if linear glycans are derivatized, or occur naturally derivatized, as does guar gum (Chapter 9), which has single-unit glycosyl branches along a backbone chain, segments are prevented from association and stable solutions result.

Stable solutions are also formed if the linear chains contain charged groups, which have coulombic repulsions that prevent segments from approaching each other. As already mentioned, charge repulsion also effects extension of chains to provide high viscosities. Such highly viscous, stable solutions are seen with sodium alginate (Chapter 12), in which each glycosyl unit is a uronic acid unit having a carboxylic acid group in the salt form. However, if the pH of an alginate solution is lowered to 3 (where ionization of carboxylic acid groups is somewhat repressed), because the pk_a values of the constituent monomers are 3.38 and 3.65, the resulting less-ionic molecules can associate to precipitate or form a gel, as expected for linear, neutral glycans (see Chapter 12).

Carrageenans are mixtures of linear chains that have a negative charge due to numerous ionized sulfate half-ester groups along the chain. These molecules do not precipitate at low pH because the sulfate group remains ionized at all practical pH values (see Chapter 11).

Gels

Gels are usually produced by direct intermolecular collisions and binding of short segments of otherwise soluble polysaccharide chains (see also Chapter 5). This formed junction has a stability that is a function mainly of its length, that is, the number of intermolecular bonds that develop. These bonds may be of various kinds (see Chapter 5). Enlargement of the junction is induced by movement of the chains that allow adjacent segments of the molecules to align. Thus, the junction grows in a zippering fashion. Further binding of segments is aided by restricting water available for solvation of the polysaccharide molecules. Reduced solvation can be effected by lowering the temperature or by addition of salts, sugars, or other substances that become highly hydrated and thus limit the number of water molecules available to solvate the polysaccharide molecules. As junction zones develop, they lead to a three-dimensional network throughout the system. The degree of overlap of chains in the junction zone determines the strength of the gel, with increased overlap increasing intermolecular binding and strength. An increase in junction zone lengths results in the three-dimensional network becoming more compact. In some gels, this results in some water being squeezed out of the gel in the form of droplets that appear on the gel surface. This process is called *syneresis*.

Gels are produced by normally soluble polysaccharides if junction zone formation is restricted to rather short chain segments. Guar gum (guaran) is a chain of β-D-mannopyranosyl units with single α-D-galactopyranosyl units on about every second chain member on average (Chapter 9). This structure forms stable solutions because the side-chain galactosyl units along the main chain prevent the intermolecular association necessary for junction zone formation. Locust bean gum has a similar structure, but with its α-D-galactopyranosyl units clumped together, leaving sections about 80 units long of naked (bare) mannan chain with no attached branches (Chapter 9). Thus, when locust bean gum is dissolved in hot water and cooled, naked sections of chains bind together by hydrogen bonding, giving rise to a weak gel. If some of the α-D-galactopyranosyl units are removed by use of an enzyme so as to lengthen the areas available for junction zone formation, the stiffness of the gel increases.

Sometimes mixed junction zones form from association of naked sections of different polysaccharide molecules. Such synergistic interaction either increases viscosity through an increase in total molecular size or, if the increase in the number of junction zones is sufficient, results in gel formation.

Gel structures can also be produced by chemically cross-linking glycan chains through the use of difunctional reagents. Poly(uronic acid) molecules can produce gels with divalent or polyvalent cations, such as calcium ions, which bind individual chains together, as described in Chapters 12 and 13.

Modification

Native polysaccharide structures can be modified, primarily by derivatization and depolymerization. Oxidation and transglycosylation (rearrangement of glycosidic linkages brought about by heat and an acid catalyst under dry conditions) may also be used. Modification changes their properties and improves or extends their functionalities.

Derivatization

Hydroxyl groups of polysaccharide molecules can be etherified and esterified just as hydroxyl groups of monosaccharides can (Chapter 2). Both ethers and esters of polysaccharides are made industrially. Ethers are the more common derivatives used to modify polysaccharides for food use.

The average number of hydroxyl groups per glycosyl unit that have been derivatized by etherification or esterification is called the *degree of substitution* (DS). A polysaccharide that contains only hexosyl units that are neither uronic acid nor deoxy-hexosyl units contains an average of three hydroxyl groups available for derivatization (substitution) per monosaccharide unit. (When a glycosyl unit is a branch point, there is one less hydroxyl group for every branching unit attached to that unit. However, every branch, even if it is a single-unit branch, is terminated with a nonreducing end-unit that contains four unsubstituted hydroxyl groups, thus balancing the one lost through branching. So, no matter how highly branched it is, a branched kind of polysaccharide [composed only of regular, neutral hexosyl units] has an average number of three hydroxyl groups per glycosyl unit.) Thus, a maximum degree of hydroxyl group substitution of three (a maximum DS of 3) can be obtained.

When hydroxyethyl ($-CH_2-CH_2OH$) and hydroxypropyl ($-CH_2-CHOH-CH_3$) ether groups are added by reaction with ethylene and propylene oxide, respectively, the substituent group itself contains a hydroxyl group that can react with another derivatizing reagent molecule. As a result, substituent chains can form and lengthen continuously. In this case, the term *molar substitution* (MS) is used to denote the average number of moles of substituent added to a glycosyl unit. Of polysaccharide ethers, only hydroxypropylstarch (Chapter 6), hydroxypropylcellulose, and hydroxypropylmethylcellulose (Chapter 7), are approved for use in food products.

Modified polysaccharides are characterized by molecular weight and physical properties as well as by DS (or MS). The actual distribution of substituents among the three hydroxyl groups available for reaction and along the polymer chain can vary with reaction conditions, reagent type, and the extent of substitution.

Depolymerization

Polysaccharides are relatively less stable to chain cleavage than are proteins and may, at times, undergo depolymerization under food processing or storage conditions. Often in the preparation of a food gum product, depolymerization is induced deliberately in order to make an ingredient that can be used at higher solids content to provide body and/or texture without producing excessive viscosity. Two types of depolymerization reactions are employed: acid-catalyzed hydrolysis and alkali-induced beta-elimination following oxidation.

Hydrolysis

Hydrolysis of glycosidic bonds joining monosaccharide (glycosyl) units in oligo- and polysaccharides may be catalyzed by either acids (H^+) or enzymes. In enzyme-catalyzed hydrolysis, the enzyme is the proton (H^+)-donating reagent. The basis for this reaction was presented in Chapter 1.

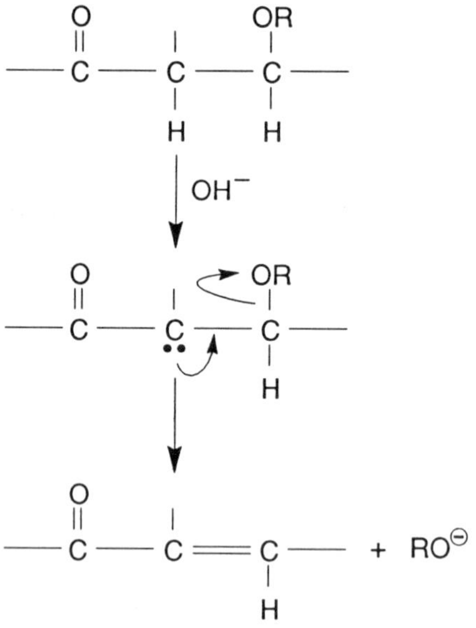

Fig. 4.7. General scheme of a beta-elimination process occurring upon removal of an acidic alpha (adjacent) proton.

The extent of depolymerization, which reduces viscosity at a given concentration, is determined by the following factors:

1. pH. The lower the pH, the faster the rate of hydrolysis/depolymerization at any given temperature,
2. The time and temperature used in the process,
3. The nature of the glycosidic linkages (monosaccharide units, anomeric configurations, linkage positions) of the polysaccharide.

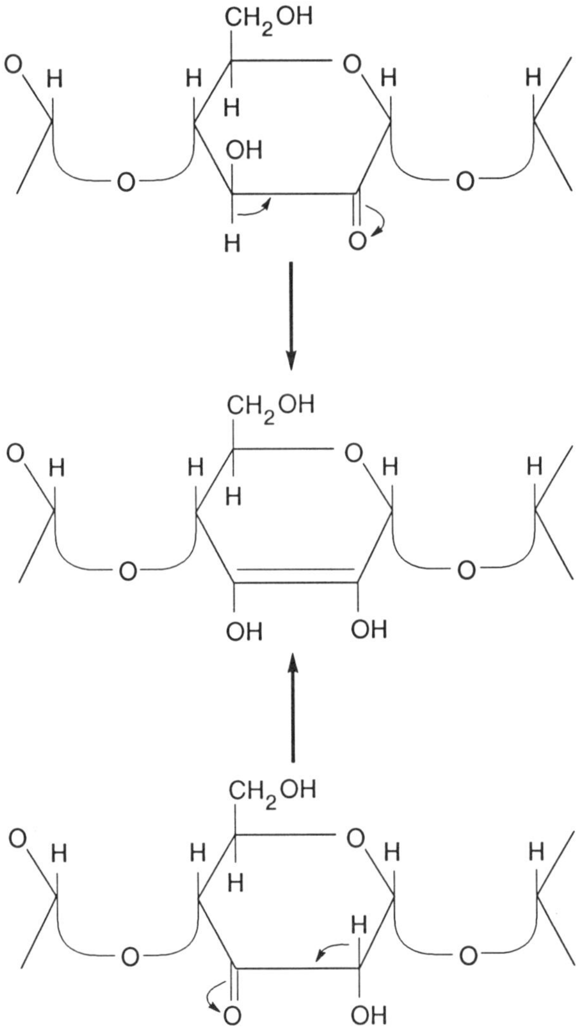

Fig. 4.8. Base-catalyzed interconversions of C-2 and C-3 carbonyl (ketone) groups in a (1→4)-linked α-glucan.

These factors are important during thermal processing, because many foods are slightly acidic, but they can also be important in determining shelf life. Loss of viscosity during processing can usually be overcome by using more of the polysaccharide (gum) in the formulation to compensate for breakdown, by using a higher viscosity-grade of the gum (Chapter 5), or by using a more acid-stable gum. Polysaccharides can also undergo depolymerization while stored as a semidry powder when hydrogen ions (H^+) are present. They are generally relatively less stable to hydrolytic breakdown than are proteins.

Polysaccharides are also subject to enzyme-catalyzed hydrolysis. The rates and end-products of breakdown are also under the control of several factors:

1. The specificity of the enzyme. Each hydrolytic enzyme is specific for a particular monosaccharide unit, its anomeric configuration, the position of the linkage—that is, (1→3), (1→4), (1→6), etc.—and its position within a polymer chain, including the nature of neighboring units.

2. pH. There is an optimum pH, usually below pH 7, for each polysaccharide-degrading enzyme.

3. Time and temperature. The rate of enzyme-catalyzed hydrolysis increases with increasing temperature up to the temperature of denaturation of the enzyme.

4. Other environmental factors. Enzyme activity and stability may be affected by factors such as salt concentration.

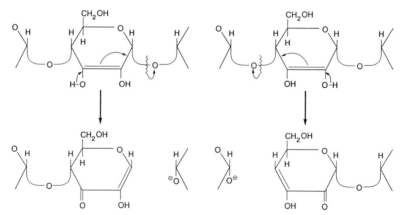

Fig. 4.9. Two mechanisms of chain cleavage via beta-elimination after oxidation of the (1→4)-α-glucan at C-2 or C-3.

Related to their susceptibility to enzyme-catalyzed hydrolysis, polysaccharides, like other carbohydrates, are subject to microbial attack. Gum products may not be delivered sterile. Thus, food products containing a polysaccharide as an added ingredient should be sterilized. Preservatives, such as 0.1% sodium benzoate, 0.17% methyl *p*-hydroxybenzoate, 0.03% propyl *p*-hydroxybenzoate, or 0.7% sorbic acid, are often added.

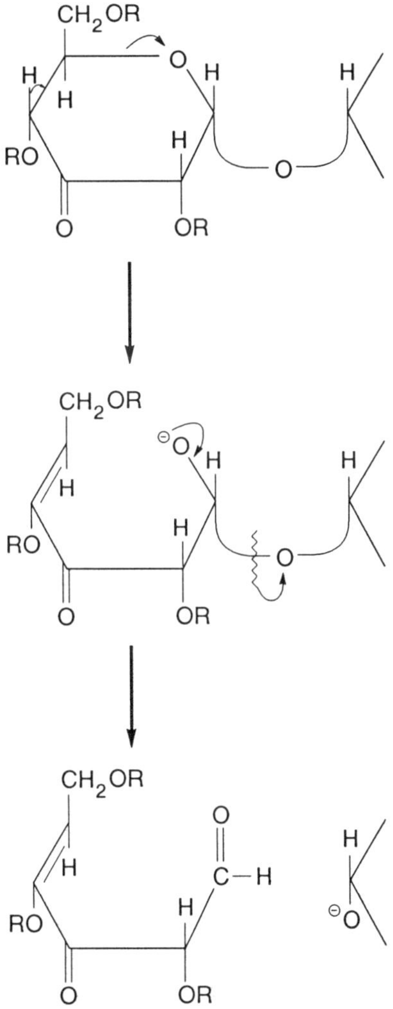

Fig. 4.10. An alternative mechanism for chain cleavage via beta-elimination after oxidation at C-3.

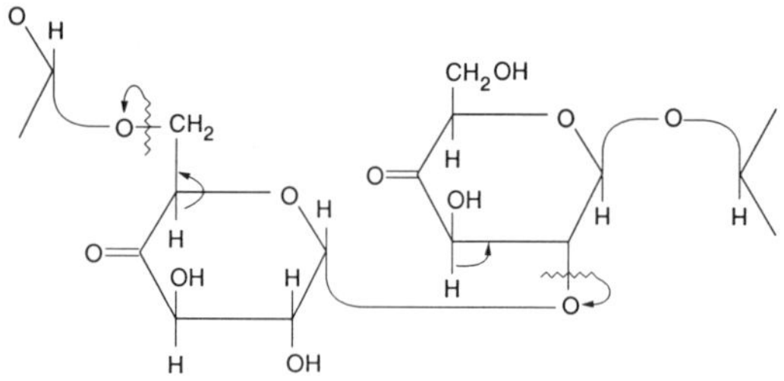

Fig. 4.11. Mechanisms of chain cleavage via beta-elimination after oxidation of C-4 in a glucan containing (1→6) and/or (1→2) linkages.

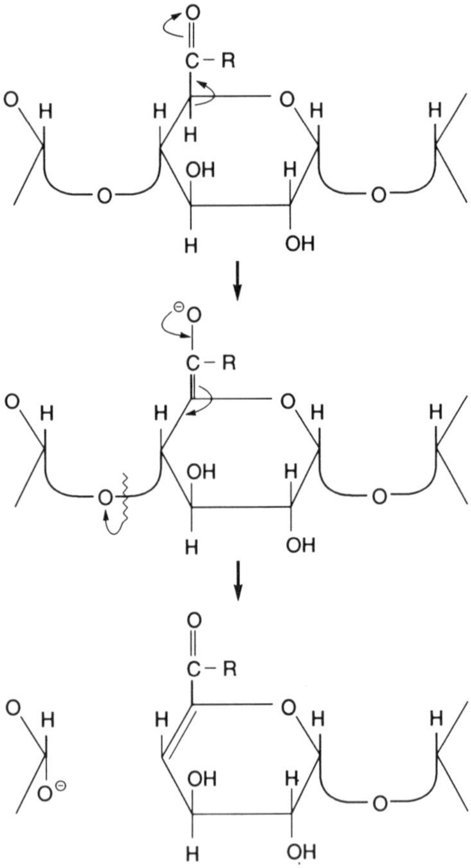

Fig. 4.12. Chain cleavage in a (1→4)-linked glucan via beta-elimination after oxidation of C-6. C-6 may be either an aldehyde or an ester group (see Fig. 13.4), both of which are electron withdrawing, making the proton on C-5 acidic, but not a carboxylate group, which is not sufficiently electron withdrawing.

Oxidation-Elimination

Polysaccharides can undergo oxidation, which can convert any hydroxyl group into a carbonyl (aldehyde or ketone) group. This oxidation is a radical process, generally involving oxygen itself and catalysis by a transition metal ion. Intermediate radicals formed from other molecules, such as unsaturated fatty acids, may also be involved. These reactions may produce other even more undesirable defects such as off-flavors and aromas. Different mechanisms for chain cleavage via beta-elimination are shown in Figures 4.7–4.12.

Depolymerization via elimination may be important in the use of poly(uronic acids); see pectin in Chapter 13.

As mentioned earlier, depolymerization is often effected deliberately by gum producers to make a variety of lower-viscosity products (Chapter 5). In making cellulose derivatives (Chapter 7), the polymer, steeped in a strongly alkaline solution, is "aged" in a manner that promotes oxidation of hydroxyl groups by dissolved oxygen. Subsequent alkali-induced base-catalyzed eliminations cleave the chain. Acids and enzymes are used to produce a variety of depolymerized starch products (Chapter 6).

Chapter 5

Behaviors of Polysaccharide Solutions, Dispersions, and Gels

Polysaccharides are used primarily to thicken and/or gel aqueous solutions and to otherwise modify and/or control rheological properties (Table 5.1). Those, other than starches, that exhibit these properties and are used commercially are called *gums* or *hydrocolloids*. Gums are generally used in food products as thickeners at concentrations of 0.25–0.50%, which indicates their great ability to produce viscosity and to form gels.

Rheology is the science of flow (liquids) and deformation (gels and other semisolids). Rheological properties are the mechanical properties of materials that flow (liquids) and deform (solids) and are expressed in terms of stress, strain, and time effects. Stress is the intensity of the force components acting on a body and is expressed in units of force per unit area. Strain is the change in size or shape of a body in response to the applied force; it is a nondimensional parameter expressed either as a ratio or as the percent change in relation to the original size or shape. Time is important in describing the rate of strain or stress, the behavior of a material under constant stress or strain, and the rate of return to the original state when the stress is removed.

Most foods are semisolids that contain some characteristics of solids and some of liquids. They are described as being *viscoelastic* because they behave as a combination of viscous and elastic elements. A material may deform in three ways; these are referred to as elastic, plastic, and viscous behaviors. In an ideal elastic material, deformation occurs the moment stress is applied, is directly proportional to the stress, and disappears instantly and completely when the stress is removed. In an ideal plastic material, deformation does not begin until a certain amount of stress, called *yield stress*, is reached; deformation is permanent, and no recovery occurs when the stress is removed. In an ideal viscous material, deformation occurs the moment stress is applied but is proportional to the rate of strain, and there is no recovery when the stress is removed.

Liquids

A common way of characterizing liquids is by measurement of their viscosity. Viscosity is the resistance to flow, that is, the resistance to applied force. It is also the shear stress divided by the shear rate. Shear stress is the applied force (pouring, mixing, pumping, chewing, swallowing, etc.). Shear rate is a value expressing how fast the liquid flows.

TABLE 5.1
General Classification of Some Major Commercial Polysaccharides According to Their Ability to Modify the Flow Characteristics of Water

Gums that form gels
  Algins
  Carrageenans
  Gellan
  Gum arabics
  Methylcelluloses
  Hydroxypropylmethylcelluloses
  Pectins
  Starches, including modified starches
  Locust bean gum + kappa-type carrageenans
  Locust bean gum + xanthan

Gums used as thickeners and stabilizers[a]
  Hydroxypropylalginates
  Carboxymethylcelluloses (certain types)
  Hydroxypropylmethylcelluloses
  Modified starches
  Xanthan

Gums used primarily as thickeners
  Carboxymethylcelluloses (certain types)
  Guar gum
  Locust bean gum

[a] Because their solutions exhibit yield stress values and/or Newtonian behavior at low shear rates.

$$\text{Apparent viscosity} = \frac{\text{unit force}}{\text{unit flow}} = \frac{\text{shear stress}}{\text{shear rate}} = \frac{\text{stress (dynes/cm}^2\text{)}}{\text{unit flow}}$$

The units of food solution viscosity are centipoises, i.e., millipascals × seconds (1 centipoise = 1 cp = 1 mPa·sec = the viscosity of water at room temperature). Viscosity is a measure of internal friction. Dissolved substances, especially polymers, increase viscosity. There are several ways to present viscosity graphically. One is to plot shear stress vs. shear rate (Fig. 5.1).

Flow properties and viscosities of liquids can be determined using one or more of a variety of instruments called *viscometers* and *rheometers*. These instruments generally fall into one of the following categories: oscillatory, rotational, tube (capillary, orifice, pipe), falling ball, sliding plate, and rising bubble. The viscosities and rheological properties of gum dispersions are often measured with

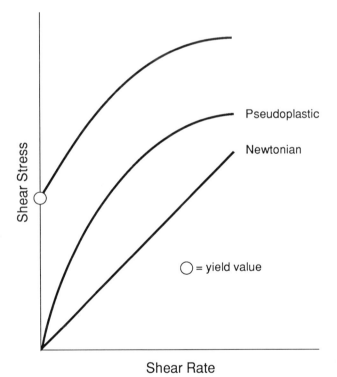

Fig. 5.1. Idealized flow curves of three different rheological systems: Newtonian flow (seldom exhibited by polymers), pseudoplastic flow, and pseudoplastic flow with a yield stress value.

rotational viscometers that measure torque (the resistance to a spindle or cylinder rotating at a given speed in a fluid). Shear rates (spindle speeds) can be changed, so one can obtain both readings at a given shear rate and plots of shear stress vs. shear rate. The latter feature is a must for polysaccharide solutions (see below). (Aqueous solutions of starches and gums often contain aggregates of hydrated molecules in addition to individual hydrated molecules and therefore are not complete molecular dispersions [true solutions].)

Starch dispersions are generally called *pastes*. The viscosity of starch pastes is most often determined with a Brabender Visco/amylo/graph. This instrument measures viscosity while starch suspensions at concentrations similar to those used in applications are cooked to form pastes, while the pastes are held hot, and while the pastes are then cooled. Its use is described in Chapter 6.

Some gum solutions appear to have a yield value, which is the amount of force that must be applied before the solution begins to flow (see Fig. 5.1). This is an important property in suspension and emulsion stabilization. To stabilize a suspension, the resistance to movement (flow) must be greater than the downward force (gravity). To stabilize an oil-in-water emulsion, the resistance to flow must prevent the oil droplets from rising to the surface. An example of a fluid with a yield value is ketchup, which requires applied force to initiate flow. Yield values of gum dispersions are the result of structural elements formed by molecular interactions that rupture when sufficient force is applied.

Types of Flow

In Newtonian flow, the rate of shear is directly proportional to the applied force. In Newtonian fluids, viscosity is independent of both shear rate and time (Fig. 5.2). Few gum solutions are Newtonian. Solutions of low-molecular-weight carbohydrates and other small molecules that form true solutions are Newtonian. Solutions of gum arabic and of low-viscosity grades of other gums at low shear rates can be essentially Newtonian.

Non-Newtonian flow is characteristic of dispersions of polymers. The viscosity of non-Newtonian liquids varies with the rate of shear, but not in a directly proportional manner. The flow behavior of non-Newtonian dispersions is determined by the properties of the particles present. The particles in a gum solution are hydrated molecules and aggregates of molecules that vary in shape, size, flexibility (that is, ease of deformation), degree of hydration, and the presence

POLYSACCHARIDE SOLUTIONS, DISPERSIONS, GELS / 95

and magnitude of charges, all properties that influence flow behavior. There are two general kinds of non-Newtonian flow of greatest importance in food systems: pseudoplastic and thixotropic.

Pseudoplastic liquids are shear-thinning liquids. In pseudoplastic flow, an increase in shear stress results in more rapid flow. This change is independent of time; that is, the rate of flow changes instantaneously as the shear rate is changed (Fig. 5.3). Because shear rate and shear stress do not vary in a linear manner, and because viscosity is the shear stress divided by the rate of shear, if meaning

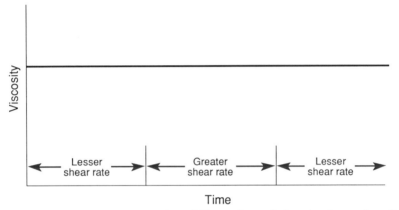

Fig. 5.2. Idealized representation of Newtonian solution rheology: viscosity as a function of shear rate and time.

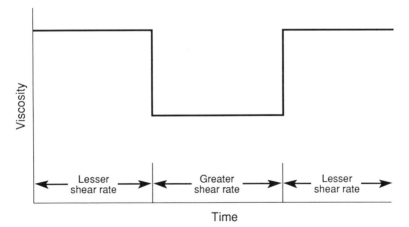

Fig. 5.3. Idealized representation of pseudoplastic solution rheology: viscosity as a function of shear rate and time.

is to be given to determined values, or comparisons are to be made between them, viscosity measurements must be made at rates of shear that are the same as those used in practice, and the shear rate (spindle speed) must be specified when reporting a viscosity value.

Pseudoplastic flow is a characteristic of linear polymers. In general, the stiffer (more rodlike) the polymer, the more pseudoplastic its solutions. Solutions of polymers that impart pseudoplastic rheology may have a significant apparent yield value. With this combination of properties, there is no flow until a force is applied, and the at-rest viscosity returns immediately upon cessation of shear, a requirement for suspensions and emulsions that are to be poured, pumped, and/or swallowed. Such shear thinning dispersions, solutions, suspensions, and emulsions require significantly less energy for stirring, pumping, and mixing. Shear thinning is nonlinear, having a log-log relationship (Fig. 5.4). Pseudoplastic gum solutions exhibit their pseudoplasticity only over a certain range of shear rates.

The degree of pseudoplasticity of a gum solution is dependent upon the concentration of the gum, its salt form if it is anionic, and its molecular weight. Thus, a gum dispersion may have almost Newtonian flow at low concentration and pseudoplastic flow after the "break point" in concentration is reached (Fig. 5.5). In general,

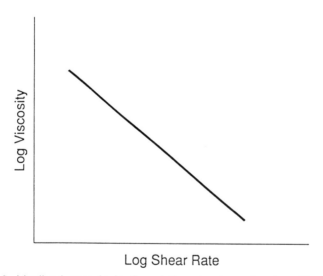

Fig. 5.4. Idealized pseudoplastic solution rheology: the logarithm of the viscosity as a function of the logarithm of the shear rate.

high-molecular-weight gums are more pseudoplastic and are, therefore, more affected by shear.

Flow properties of liquids influence their mouthfeel, as well as their processability. A slimy material is one that is thick, coats the mouth, and is difficult to swallow. Viscous gum solutions that are less pseudoplastic are said to give long flow[1] and are perceived as being slimy. More pseudoplastic viscous solutions are described as having short (pituitous) flow[1] and are perceived as nonslimy (Fig. 5.6). Thus, sliminess is inversely related to pseudoplasticity. To be perceived as nonslimy, the system must thin sufficiently at the low shear rates produced by chewing and swallowing so that it will clear the mouth. The degree of pseudoplasticity of some gum solutions can be modified by controlling the concentration, viscosity grade,

---

[1] *Short* and *long flow* refer to draining behavior from a pipet or funnel. As the forming drop gets larger, the force of gravity on it becomes greater, making it heavier. This causes the liquid's flow rate to increase as the mass of the drop increases. In a pseudoplastic fluid, the increased flow rate causes the viscosity and the diameter of the stream to decrease. Finally, the stream breaks. The result is that the fluid exits the pipet in "short" drops. Fluids without shear-thinning behavior do not exhibit the increased flow rate, and the stream exits the pipet in "long" strings.

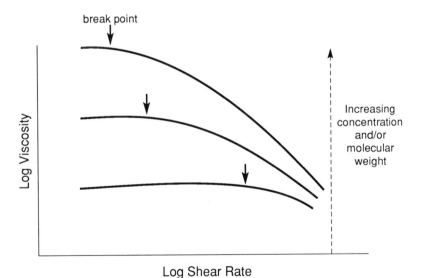

Fig. 5.5. Idealized pseudoplastic solution rheology: the logarithm of the viscosity as a function of the logarithm of the shear rate as influenced by polymer concentration and/or molecular weight.

and/or salt form of the gum or the solution pH, which determines the degree of ionization of the gum if it contains uronic acid units. Texture is also a function of tackiness, smoothness (grittiness or powderiness), plasticity, particle size, particle density, temperature, and viscosity. A slight degree of sliminess is at times desirable in that the coating left in the mouth mimics that left by a fatty food, providing a low- or no-fat food with the perception of richness or creaminess.

Processing conditions that are influenced by flow properties include such characteristics as energy required for mixing and pumping, rate of flow through pipes, and behavior during

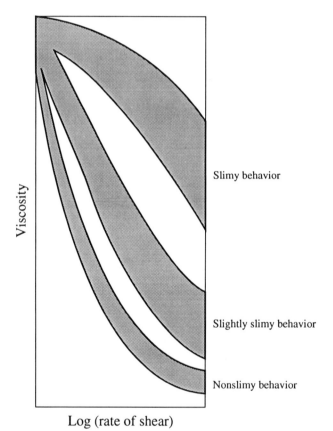

Fig. 5.6. General relationship of the degree of pseudoplasticity to the perception of sliminess. (Adapted from A. S. Szczesniak and E. Farkas, *J. Food Sci.* 27:381-385, 1962)

pasteurization and on scraped-surface heat-exchangers and in any other operation in which an applied force causes flow.

Thixotropic flow is another type of non-Newtonian flow. The viscosity of thixotropic solutions, like that of pseudoplastic solutions, decreases as the rate of shear increases, but in a time-

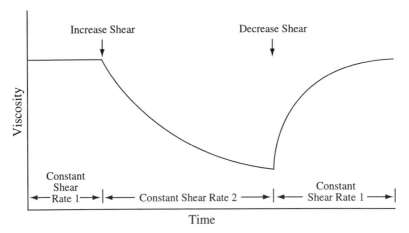

Fig. 5.7. Idealized representation of thixotropic rheology, showing the time dependence of the change in viscosity with a change in shear rate. Compare with pseudoplastic rheology (Fig. 5.3).

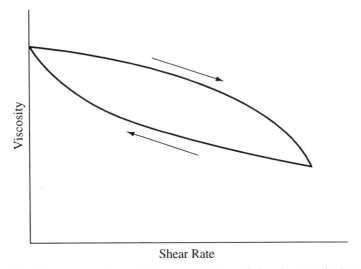

Fig. 5.8. One type of graphical presentation of the hysteresis loop of thixotropic solutions: viscosity versus shear rate.

dependent manner rather than instantaneously. It also regains its original level after cessation of shear, but again only after a measurable time interval rather than instantaneously (Fig. 5.7). This behavior is due to a gel → sol (fluid molecular dispersion) → gel transition. A thixotropic solution at rest is a weak (pourable) gel.

Solutions of cellulose derivatives may exhibit thixotropic behavior. The degree of thixotropy is a function of the degree of substitution (DS), uniformity of substitution, and molecular weight of the polymer (see Chapter 7). Cellulose molecules associate easily and strongly with each other (see Chapter 4), so nonuniformly derivatized cellulose molecules that have stretches of unmodified β-D-glucopyranosyl units can form limited intermolecular aggregates, called *junction zones* (see section on gels in this chapter). These interactions are responsible for the weak, easily broken gel structure. Energy may be required to break the junction zones and start flow. Hysteresis loops are characteristic of thixotropic flow (Fig. 5.8). Pulp in fruit juices provides a slight thixotropic characteristic. A dry mix for a fruit drink should mimic this behavior.

Viscosity Grades of Gums

Industrial gums vary in degrees of pseudoplasticity and/or thixotropy. Most gums are available in a wide range of viscosity grades

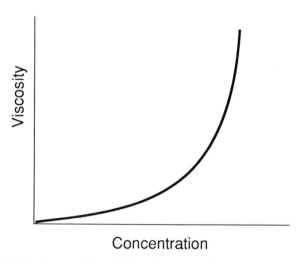

Fig. 5.9. Idealized curve of the relationship of viscosity to concentration of a gum in water. At "higher" concentrations, which are still generally low (less than 5%), random coils of polymer begin to overlap one another to form an entangled network, giving a much greater concentration dependence to the viscosity.

(Figs. 5.9 and 5.10), different viscosity grades being provided for different specific applications.

It is almost as important to select the proper viscosity grade of a gum as it is to select the most efficacious gum for a particular application. If viscosity is the attribute desired, a high-viscosity gum at a low solids concentration should be used. If binding, stabilization, or coating, for example, is the goal, a low-viscosity gum at a high solids concentration should be used. Several viscosity grades of a given gum should be tried for each application in order to find the one that works best. Comparisons should be made since viscosity is a function of the degree of hydration of gum particles, that is, the degree of molecular dispersion, which in turn is determined by the conditions of dissolution/dispersion. The manufacturer's data or the data of others can be used for comparisons of the various viscosity grades of a single product, but comparisons of different gums should be made by the user.

The viscosity of gum solutions generally increases approximately logarithmically with concentration (doubling the concentration gives about a 10-fold increase in viscosity).

Gun Dissolution

Gums "dissolve," that is, form molecular dispersions, by particle swelling, followed by disruption of the swollen particles (Fig. 5.11).

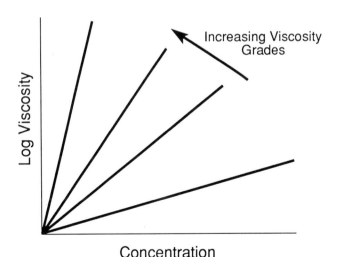

Fig. 5.10. Idealized curves of the logarithm of the viscosity as a function of gum concentration for four viscosity grades of a single gum.

The rate of hydration can be controlled. Degree and rate of dissolution are influenced by the 1) type of gum, 2) viscosity grade, 3) particle size, 4) temperature, 5) pH, 6) presence of other solutes, 7) type and amount of counterions (if an ionic gum), 8) means of dispersion, and 9) any surface treatment given particles.

In the preparation of molecular dispersions of gums, several factors are important (see also Chapter 4).

1. Dispersion of gum particles. Particles must be well dispersed in the aqueous medium before swelling begins. Besides employing good agitation, the user may premix the gum with another ingredient, usually sugar or an oil.

2. Agitation of liquid. Several special pieces of equipment are available for thorough agitation to disperse gum particles rapidly.

3. Composition of hydration medium. It is best to dissolve the gum in water first and then add other ingredients.

4. Particle size (Fig. 5.12). A fine-mesh gum is likely to lead to clumping and has a greater requirement for good dispersion before swelling. For dispersion of a dry mix containing a gum in the home kitchen, it is best to use coarse particles.

5. Nature of the gum itself.

6. Temperature.

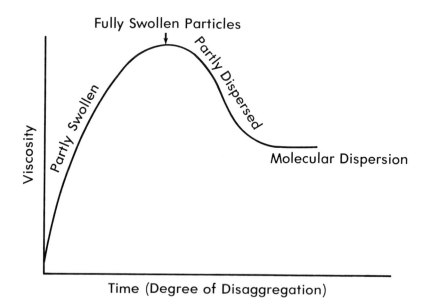

Fig. 5.11. Idealized curve of the development of viscosity with time as gum particles swell and disintegrate in a stirred (sheared) aqueous system.

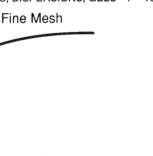

Fig. 5.12. Influence of particle size on the development of viscosity with time/degree of hydration.

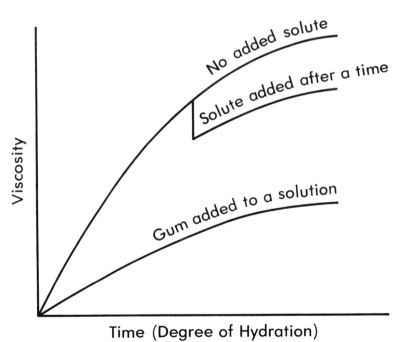

Fig. 5.13. Effect of solutes added before or after the gum is added to water on the development of solution viscosity as a function of time (compare Fig. 5.12).

In general, ionic gum molecules are more soluble and dissolve faster than neutral gum molecules of the same size and shape, but the solubility of ionic gums is much more depressed by salts than is the solubility of neutral gums. It is unlikely that full viscosity will be obtained if a gum, especially an ionic gum, is added to a salt/solute solution (Fig. 5.13).

Other Gum Solution Behavior

Most gum solutions decrease in viscosity as temperature is increased (Fig. 5.14). Xanthan is an exception between 0 and 100°C (see Chapter 10). Often, therefore, gum solutions are made at an elevated temperature where solubility is high and viscosity is low, then cooled for thickening. Prolonged heating may produce degradation of the gum, the degree of which is controlled by the temperature, pH, and inherent stability of the polysaccharide (Fig. 5.15). When viscosity loss occurs, more gum may be added initially. With locust bean gum, some heating is necessary to obtain full viscosity (Fig. 5.16), so viscosity may increase during processing of food formulations containing it. Solutions of methyl- and hydroxypropylmethylcelluloses first thin when heated and then gel at particular temperatures (Chapter 7).

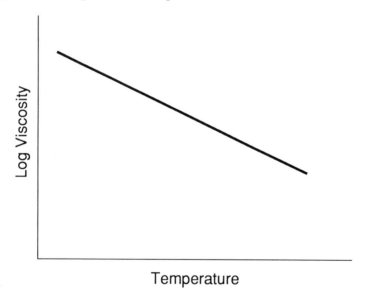

Fig. 5.14. Idealized relationship of the logarithm of viscosity to solution temperature. For most gums, this relationship is reversible. The slope of the log viscosity versus temperature line is different for each gum.

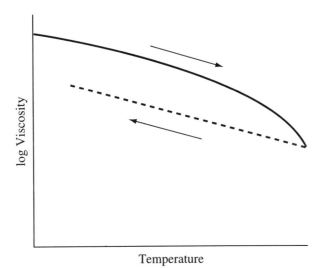

Fig. 5.15. Idealized relationship of the logarithm of viscosity to solution temperature for a gum that is inherently unstable to heat at the solution pH (compare Fig. 5.14). Note that the solution increases in viscosity as it is cooled (dotted line).

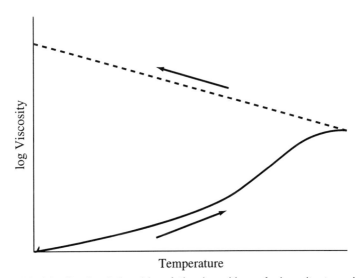

Fig. 5.16. Idealized relationship of the logarithm of viscosity to solution temperature for a gum that requires heating to achieve full hydration and dispersion. Note that the solution increases in viscosity as it is cooled (dotted line).

The viscosity of solutions of ionic gums[2] is affected by pH. Usually, the viscosity increases markedly below pH 2.5–5.0 (Fig. 5.17). Some anionic gums become insoluble at low pH; solutions of some gel at low pH. The same gum may either precipitate or bring about gelation, depending on its concentration and how the pH is adjusted. The pH of a gum solution should be adjusted to an acidic pH only after the gum is dissolved.

The viscosity of gum dispersions is affected by the presence of substances that compete for water. Salts and other solutes decrease gum hydration. With some gum solutions, salts/solutes increase viscosity or effect gelation, probably by increasing intermolecular interactions. With other gum solutions, salts/solutes decrease viscosity, probably by increasing intramolecular interactions, that is, causing more molecular coiling. Polyols, including sugars, often increase viscosity.

Viscosities of gum solutions are also influenced by interactions

---

[2] All food gums are either neutral or anionic. Hence, ionic food gums are anionic polymers, most often containing carboxyl (-COOH/-COO⁻) groups.

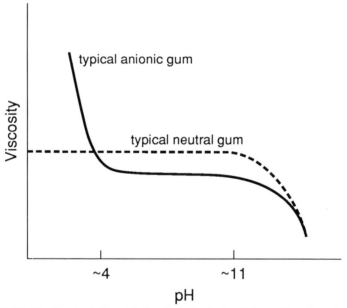

Fig. 5.17. Idealized relationship for the effect of solution pH on the viscosity of solutions of neutral gums and some gums with carboxylic acid groups (uronic acid groups). Alkaline pH values (above pH 8) are rarely, if ever, encountered in foods.

with other polymers. Some synergetic relationships, in which the viscosity or gel strength of the combination is greater than that predicted by summing the individual properties, are given in Table 5.2.

Gels

A gel is a continuous, three-dimensional network of connected molecules or particles (such as crystals, emulsion droplets, or molecular aggregates/fibrils) entrapping a large volume of a continuous liquid phase (Fig. 5.18). In many food products, the gel network consists of polymer (polysaccharide and/or protein) molecules or fibrils formed from polymer molecules joined in junction zones by hydrogen bonding, hydrophobic associations (that is, van der Waals attractions), ionic cross-bridges, entanglements, or covalent bonds, and the liquid phase is an aqueous solution of low-molecular-weight solutes and portions of polymer chains.

Gels have some characteristics of solids and some characteristics of liquids. When polysaccharide molecules or fibrils formed from polymer molecules interact over portions of their lengths, forming junction zones and a three-dimensional network, a fluid solution is transformed into a solid, spongelike structure that can retain its shape to some degree. The three-dimensional network structure provides resistance to applied stress, causing it to behave somewhat as an elastic solid. However, the continuous liquid phase, in which molecules are completely mobile, makes a gel less stiff than an ordinary solid, causing it to behave somewhat as a viscous liquid. Therefore, a gel is a viscoelastic semisolid; that is, the response of a gel to stress is partly characteristic of an elastic solid and partly character-

TABLE 5.2
Synergisms

Viscosity-enhancing combinations
Carboxymethylcelluloses + guar gum
Carboxymethylcelluloses + casein
Carboxymethylcelluloses + soy protein
Kappa-type carrageenans + kappa-casein
Xanthan + agarose
Xanthan + κ-type carrageenans
Xanthan + guar gum
Gelling combinations
Kappa-type carrageenans + locust bean gum
Xanthan + locust bean gum

istic of a viscous liquid. Hence, gels exhibit some of the behaviors of elastic solids and some of the behaviors of polymer solutions.

Junction zones, formed by intermolecular interactions, can be any one of several different types:

1. Intermolecular interactions between regular, linear segments of molecules of the same polysaccharide (for example, methylcelluloses and hydroxypropylmethylcelluloses [Chapter 7], high-DM pectin[3] [Chapter 13], and starches [Chapter 6])

2. Intermolecular interactions between regular, linear segments of polyanionic polysaccharide molecules effected by a cation (for example, alginates [Chapter 12], low-DM pectin [Chapter 13], carrageenans [Chapter 11], and gellan)

3. Interactions between linear chain segments of two different polysaccharide molecules (for example, locust bean gum + carrageenan [Chapter 11] and locust bean gum + xanthan [Chapter 10])

4. Interactions between ionic polysaccharide molecules and protein molecules (for example, κ-carrageenan + κ-casein [Chapter 11])

---

[3] DM = degree of methylation/esterification (see Chapter 13).

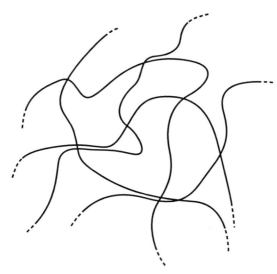

Fig. 5.18. Diagrammatic representation of the type of three-dimensional network structure found in gels. This type of structure is known as a fringed micelle structure. Parallel side-by-side chains indicate the ordered, crystalline structure of a junction zone. The holes between contain an aqueous solution of dissolved segments of polymer chains and other solutes.

5. Crosslinking of neutral polysaccharide molecules with a multivalent anion (a form of gelation not used in foods)

6. Chain entanglements, especially in the case of branch-on-branch polymers (for example, gum arabic [Chapter 14] and starch amylopectin [Chapter 6]).

As a general rule, regular, linear segments of polysaccharide molecules are in the form of a helix. Intermolecular interactions result either in simple associations or in the formation of a double (or, in some cases, a triple) helix. The relatively stiff, linear, double helical segments may then interact (pack together) to form a super junction and the three-dimensional gel network.

Junction zones between molecules or particles must be of limited size. If molecules or particles interact over a major portion of their length, precipitation results. Therefore, regular, linear chain segments must be interrupted by irregularities so that interactions take place over only limited segments of a molecule and a limited junction zone is formed. In general, the larger the junction zones, the more difficult it is to separate molecules by shearing forces and thermal energy. Also, in general, if junction zones grow after formation of the gel, the network becomes more compact, the structure contracts, and syneresis occurs.

Although gel-like or salve-like materials can be formed by high concentrations of particles (much like tomato paste), to form a true gel, the polymer molecules or aggregates of molecules must first be in solution, then associate in junction zone regions to form the three-dimensional gel network structure.

Polysaccharide gels generally contain only about 1% polymer, that is, they may contain 99% or more water. Such gels can be quite strong, even though they contain small concentrations of polymer. Examples are dessert gels, aspics, structured fruit pieces, structured onion rings, meat-analog pet foods, and icings. To understand why and how polysaccharide solutions can gel at such low concentrations, the relatively new gelling food gum gellan can be used as an example. (Calculations for other polysaccharides will give essentially the same results.) Consider 200 ml (a small jelly glass) of a gel made by cooling a hot 1% solution of gellan. (The gel formed will be quite rigid, like an agar gel.) X-ray fiber diffraction reveals that gellan is in the form of a parallel double helix and that the double helices pack together in an antiparallel manner in crystalline arrays. Chemistry, x-ray fiber diffraction, and modeling show that the polysaccharide contains a tetrasaccharide repeating unit with the distance from the start of one repeating unit to the start of another of 18.8 Å. Therefore, there are eight sugar units in each 18.8 Å of the double helix. Knowing the molecular weight of a tetrasaccharide repeat unit, it is easy to calculate the number of moles of repeating units in 2 g (the amount in 200 ml of a 1% solution) of gellan; it is about 3 mmol. Using Avagadro's number and the 18.8 Å length of two repeat units in the calculation discloses that the 200 ml of solution contains about 1 billion miles (1.7×10^9 km) of double helix. That is about 5,000 times the distance to the moon. Thus, there is enough length of polymer to give a substantial gel network. In this case, the junction zones are probably formed between fibrils formed by the packing together of double helical structures held in place by cations, as depicted in Figure 5.19.

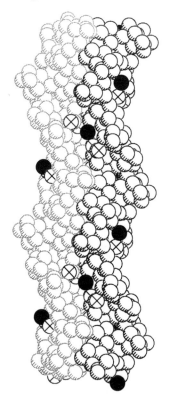

Fig. 5.19. Double helix of native gellan. Light and dark shading distinguishes the two chains. Potassium ions forming salt bridges holding the chains together are shown as black circles. Circles with crosses are the oxygen atoms of carboxylate groups. (Courtesy of R. Chandrasekaran and A. Radha)

Other solutes that compete for water increase interactions between gum molecules and thereby affect the texture of the gel. In the case of high-DM pectin (Chapter 13), a high sugar concentration is required for gelation.

Gums that form gels by intermolecular interactions between uniform, linear segments of the same polysaccharide are high-DM pectins, amylose, alginic acid, and methyl- and hydroxypropylmethylcelluloses (Chapters 13, 6, 12, and 7, respectively). The thixotropy of solutions of the high-degree-of-polymerization, low-degree-of-substitution carboxymethylcelluloses (CMC) is due to the same mechanism (Chapter 7).

κ-Type carrageenans (Chapter 11) and gellan form gels with potassium ions. Low-DM pectins, alginates, iota-type carrageenans, certain CMCs (Chapters 13, 12, 11, and 7, respectively), and gellan form gels with di- and trivalent cations.

Examples of gums that form gels by interactions between linear chain segments of two different polysaccharide molecules are locust bean gum (LBG) + xanthan and LBG + κ-type carrageenans (Table 5.3 and Chapters 9–11).

A soft, thixotropic gel forms when κ-carrageenan, an anionic gum, interacts with the κ-casein of milk, which takes place even though the pH of milk is above the isoelectric pH value (pI) for casein and has a net negative charge (Chapter 11). Other anionic gums (for example, CMC) will interact with proteins when the polysaccharide is in the anionic form (pH value above its pK_a) and the protein is in a cationic form (pH value below its pI).

Galactomannans (LBG and guaran, Chapter 9), which have *cis*-vicinal hydroxyl groups, form gels in the presence of borate and similar anions at alkaline pH values. This gelling mechanism finds extensive use in hydraulic fracturing for oil recovery, but borate cannot be used in foods.

Gelation of amylopectin and gum arabic solutions may result from chain entanglements. It is likely that these very large, bushlike molecules simply overlap into each other's space; ensuing entanglements form a network structure.

A texture profile analysis can be obtained using the Instron universal testing machine (Fig. 5.20). Parameters that are measured are as follows:

1. Modulus, also known as the modulus of elasticity or Young's modulus, is the initial slope of the force vs. deformation curve and is a measure of the firmness of a gel when it is compressed a small amount,

that is, before the limit of elasticity is reached and the gel ruptures. It usually correlates closely with the sensory perception of firmness.

2. Hardness is the maximum force exerted during the first compression cycle. There are often two yield points. The first inflection point indicates a failure of some structural elements in the gel. The final peak is the rupture yield and indicates a massive failure of the material; it is correlated with what is commonly called *gel strength* (rupture, strength, compressive strength).

3. Brittleness, also known as fracturability, is the percent compression (the amount of strain) when the sample breaks. The smaller the number, the more brittle the gel.

TABLE 5.3
Formation of Polysaccharide Food Gels

Gelling Conditions	Polysaccharide
With acids	Sodium alginates
	High-DM[a] pectins + sugar
With cations	Sodium alginates + Ca^{2+}
	Iota-type carrageenans + Ca^{2+}
	Low-DM pectins + Ca^{2+}
	Kappa-type carrageenans + K^+
	Carrageenans + protein (above pI)
	Carrageenans + pectin + protein (below pI)
Upon cooling of hot solutions	Agar
	Calcium alginates
	Gellan
	High-DM pectins + acid + sugar
	Calcium pectinates
	Kappa-type carrageenans + locust bean gum
	Xanthan + kappa-type carrageenans
	Xanthan + locust bean gum
	Xanthan + agarose
	Starches
Upon heating of solutions	Curdlan (irreversible)
	Hydroxypropylmethylcelluloses (thermoreversible)
	Methylcelluloses (thermoreversible)
With addition of solutes	High-DM pectins + acid
Shear-reversible gels	Iota-type carrageenans (above pH 5.0, 20–80 Brix)
	Low-DM pectins (below pH 5.0, 20–80 Brix)

[a] Degree of methylation/esterification.

4. Elasticity, also called springiness, is a measure of how much the deformed material returns to its undeformed condition after the deforming force of the first compression cycle is removed. Elasticity correlates with how rubbery a gel feels in the mouth.

5. Cohesiveness is a measure of how much the sample is broken down by the first compression cycle. It is determined by dividing the work done on the gel (that is, the energy required to compress) during the second compression cycle by the work done during the first compression cycle. Gels with high values are perceived as being tough and difficult to break up in the mouth.

Some other terms that are used to describe semisolid foods are *adhesiveness*, which is any negative force after the first compression cycle and a measure of how sticky, tacky, or gooey a material is; *chewiness*, which is a combination of hardness, cohesiveness, and elasticity; and *gumminess*, which is the energy required to disintegrate a semisolid food product to a state ready for swallowing and is a combination of hardness and cohesiveness.

The results of a texture profile analysis can be presented as in Figure 5.21.

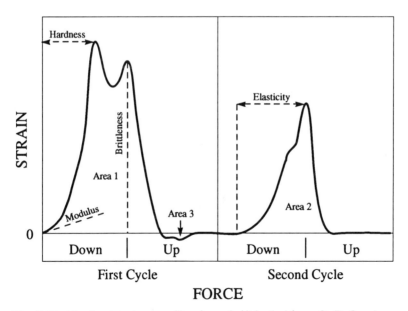

Fig. 5.20. Idealized texture profile of a gel. (Adapted from G. R. Sanderson, V. L. Bell, R. C. Clark, and D. Ortega, in *Gums and Stabilisers for the Food Industry*, G. O. Phillips, P. A. Williams, and D. J. Wedlock, eds. IRL Press, Oxford, 1988)

It is obvious that the characteristics described above are related to such things as eating quality, use properties (for example, cuttability and spreadability), and handling properties (for example, shape retention) under different conditions and time scales.

Choosing a Thickening or Gelling Agent

In choosing the best starch or gum product for a particular application, consideration of rheological characteristics of solutions or gels usually dominates. In selecting a gum for its ability to produce viscosity, the following factors must be considered:

1. Available viscosity grades
2. Amount needed of the proper viscosity grade to give a specified viscosity or gel strength
3. Cost/functionality (cost of the amount needed to impart the desired functional characteristics)
4. pH of the system (effect on hydration rates and stability)
5. Temperatures during processing and times at those temperatures (effect on hydration rates and stability)
6. Times at each processing and storage temperature (effect on hydration rates and stability)
7. Interactions with other ingredients, i.e., competition with other dissolved substances for water

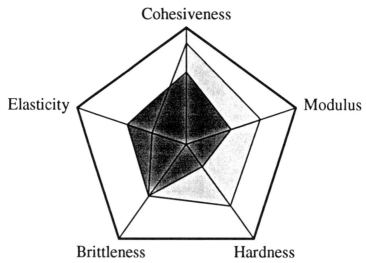

Fig. 5.21. One example (idealized) of a way to present the results of a texture profile analysis that compares gels prepared from two different gums.

8. Desired texture
9. Ease of dispersion with available equipment.

Gums may enhance flavor and sweetness in food products through their thickening and coating, resulting in the food being held in the mouth longer. However, in general, the perceived intensity of flavor and sweetness is suppressed as the viscosity increases, presumably due to reduced rates of transport of fresh flavor or sweetener to the taste buds. As with the perception of sliminess (Fig. 5.6), the more shear-thinning is the solution rheology, the less is the flavor/taste suppression.

Some characteristics of food gels that must be considered in choosing the polymer or polymers to form a gel are means of gelation, shear reversibility, heat reversibility, texture profile, gel strength, degree of syneresis, clarity, and freeze-thaw stability. Gelation of polysaccharide solutions may be caused by additives (generally acids or cations), by cooling a hot solution, or, in the case of some methylcelluloses (see Chapter 7), by heating a cool solution. Most gels are heat reversible, and gelling and melting temperatures are often important gel characteristics. Most gels are shear irreversible, but thixotropic solutions can be considered to be pourable, reversible gels. The texture of a gel may be brittle, elastic, plastic, rubbery, or tough. Its strength may be firm or soft. It may be cuttable, pourable, spoonable, or spreadable. Achieving the desired properties is a matter of formulating a proper gelling system.

When characteristics other than specific flow properties and gel formation and properties are required, consideration must be given to them also. Other properties of polysaccharides are responsible for their ability to function as adhesives, binders, bodying agents, bulking agents, crystallization inhibitors, clarifying agents, cloud agents, emulsifying agents, emulsion stabilizers, encapsulating agents, fat sparers, film formers and coating materials, flocculating agents, foam stabilizers, mold release agents, suspension stabilizers, swelling agents, syneresis inhibitors, texturing agents, and whipping agents and to participate in water absorption and binding (water retention and migration control). Storage stability, including freeze-thaw stability, is often an underlying criterion for food products. Each food starch or gum generally has one or more outstanding properties related to a functional characteristic that is the basis for its choice. Properties of specific food starches and gums are presented and discussed in subsequent chapters.

Chapter 6

Starch

Starch's unique chemical and physical characteristics and nutritional quality set it apart from all other carbohydrates. It is the predominant food reserve substance in plants and provides 70–80% of the calories consumed by humans worldwide. Starch, products derived from starch, and sucrose constitute most of the digestible carbohydrate in the human diet, lactose being the only other carbohydrate digestible by humans. The amount of starch used in the preparation of food products greatly exceeds the amount of all other food hydrocolloids combined, and this does not include the starch in flours used to make bread and other bakery products, that naturally occurring in grains used to make breakfast cereals, nor that in fruits and vegetables, such as potatoes.

Commercial starches are obtained from seeds, particularly corn, waxy corn (waxy maize), high-amylose corn, wheat, and various rices, and from tubers or roots, particularly potato, sweet potato, and cassava (tapioca starch).

Starches and modified starches have an enormous number of food uses, including adhesion, binding, clouding, dusting, film-formation, foam strengthening, antistaling, gelling, glazing, moisture retention, stabilizing, texturizing, and thickening applications. Certain modified starches with particular specific physical characteristics can even play a role in lowering the fat content of prepared foods by providing a similar sensory perception of fattiness or creaminess.

Starch is unique among carbohydrates because it occurs naturally as discrete particles, called *granules* (Fig. 6.1). Appearances of granules are so different among different plants that even amateur microscopists can identify the crop source of most commercial starches. Starch granules are relatively dense, are insoluble, and

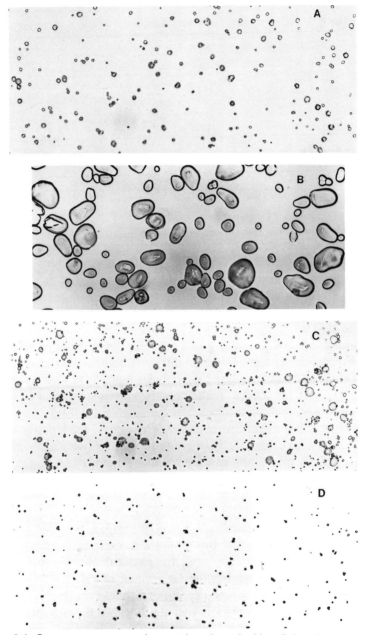

Fig. 6.1. Some common starch granules viewed with a light microscope: **A,** common corn starch; **B,** potato starch; **C,** wheat starch; **D,** rice starch. All photographs are of the same magnification.

hydrate only slightly in room-temperature water. They can be dispersed in water, producing low-viscosity slurries that can be easily mixed and pumped even at concentrations of 35% or greater. The thickening power of starch is realized only when a slurry of granules is heated. A 5% slurry of unmodified starch granules heated to about 80°C (175°F) with stirring produces very high viscosity.

A second uniqueness is that most starch granules are composed of a mixture of two polymers: an essentially linear polysaccharide called *amylose* and a highly branched polysaccharide called *amylopectin*.

Amylopectin

Amylopectin, is a very large, highly branched molecule. Starch molecules grow from a single β-D-glucopyranosyl unit attached to a self-glucosylating initiator protein molecule, amylogenin. Other D-glucopyranosyl units are added sequentially, being donated by adenosine diphosphate glucose molecules to produce a chain of α-D-glucopyranosyl units joined by (1→4) linkages (Chapter 4). However, in addition to the chain-lengthening enzyme, a branching enzyme is active. The branching enzyme seems to need a linear chain of 40–50 units before it transfers a portion of the chain, which becomes an α-D-(1→6)-linked branch, whereupon both nonreducing ends may continue to be elongated. The branch point linkages constitute 4–5% of the total linkages. An amylopectin molecule consists of a main chain, called the C chain, which carries the one reducing end-group and numerous branches, termed B chains, to which third-layer A chains are attached.[1] The branches of amylopectin molecules are clustered as shown in Figure 6.2. At least some, if not most, branches occur as double helices of parallel chains.[2]

---

[1] A chains are those that are connected to another chain via (1→6) linkages but are themselves unbranched. B chains are chains that are connected to another via (1→6) linkages and also have one or more A or other B chains attached to them via (1→6) linkages.

[2] A double helix consists of two chains coiled around a common axis. It is the structure obtained if one takes two rods made of something flexible like rubber, places them side by side, holds one end steady, and twists the other end to make helices out of each of the two rods. Deoxyribonucleic acid (DNA) is an example of a double helix of antiparallel chains.

Structures, average molecular weights, and molecular weight ranges of amylopectins vary with the botanical source. However, most amylopectin molecules have a trimodal distribution of A and B chain lengths. For potato starch, the average degree of polymerization (DP) values of these three fractions are 34–45, 23–32, and 13–15, the smallest of which consists of the outermost, that is, the A, chains. Molecular weights of from 10^7 to up to 5×10^8 make amylopectin molecules among the largest in nature.

Amylopectin is present in all known starches, constituting about 75% of most common starches (Table 6.1). Some starches consist entirely of amylopectin and are called *waxy starches*. Waxy corn (waxy maize), the first grain recognized as containing only

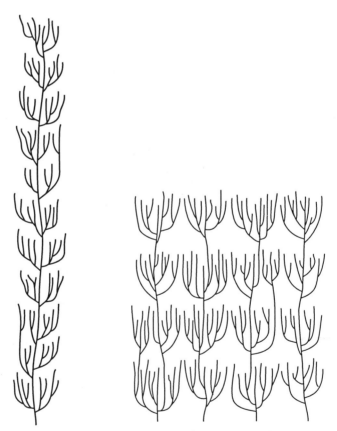

Fig. 6.2. Diagrammatic representation of a portion of an amylopectin molecule (left). Enlargement of typical packed clusters (right). Individual chains are helical and pairs of chains are double helical.

amylopectin in its starch, was so termed because, when the kernel is cut, the new surface appears vitreous or waxy, but there is no wax present. Other all-amylopectin starches are also called *waxy*.

Potato amylopectin is unique in having phosphate ester groups attached to one in about every 200–550 α-D-glucopyranosyl units. The phosphate ester groups are located near branch points, most often (60–70%) at an O-6 position, but some also at O-3 positions. About 88% are found on B chains. Due to its phosphate ester content, potato starch has a negative charge and a pK_a of 3.7. The resulting slight coulombic repulsion may contribute to the rapid swelling of potato starch granules in warm water and to the high viscosity, good clarity, and low rate of retrogradation of potato starch pastes.

Phytoglycogen, which is similar to amylopectin but has a higher degree of branching (about 10% of all linkages), is present in sweet corn in amounts of up to 25% and is water soluble.

TABLE 6.1
General Properties of Some Starch Granules and Pastes

	Corn Starch	Waxy Maize Starch	High-Amylose Corn Starch	Potato Starch	Tapioca Starch	Wheat Starch
Granule size, μm	2–30	2–30	2–24	5–100	4–35	2–55
% Amylose	28	<2	50–70	21	17	28
Gelatinization/ pasting temp. (°C)[a]	62–80	63–72	66–170[b]	58–65	52–65	52–85
Relative viscosity	Medium	Medium high	Very low[b]	Very high	High	Low
Paste rheology (body)[c]	Short	Long	Short	Very long	Long	Short
Paste clarity	Opaque	Slightly cloudy	Opaque	Translucent	Translucent	Cloudy
Tendency to gel/ retrograde	High	Very low	Very high	Medium to low	Medium	High
Lipid, % DS[d]	0.8	0.2	<0.1	<0.1	<0.1	0.9
Protein, % DS[d]	0.35	0.25	0.5	0.1	0.1	0.4
Phosphorus, % DS[d]	0.00	0.00	0.00	0.08	0.00	0.00

[a] From the initial temperature of gelatinization to complete cookout.
[b] Under ordinary cooking conditions, where the slurry is heated to 95–100°C, high-amylose corn starch produces essentially no viscosity. Cookout does not occur until the temperature reaches 160–170°C (320–340°F). However, loss of birefringence begins at about 66°C.
[c] See Chapter 5, footnote 1.
[d] DS = dry solids.

Amylose

While amylose is essentially a linear chain of (1→4)-linked α-D-glucopyranosyl units, many amylose molecules have a few α-D-(1→6) branches, perhaps 0.3–0.5% of the total linkages. The branches are generally either very long or very short and are separated by large distances, allowing the molecules to act as essentially linear polymers, forming strong films and fibers and retrograding easily (see Retrogradation and Staling). The average molecular weight of amylose molecules is about 10^6.

The axial→equatorial position coupling of the (1→4)-linked α-D-glucopyranosyl units in amylose chains (Fig. 4.6) gives the molecules a right-handed spiral or helical shape (Fig. 6.3). A consequence of helix formation is that films and fibers from amylose are more elastic than films and fibers from cellulose molecules. The interior of the helix contains predominantly hydrogen atoms and is lipophilic, while the hydroxyl groups are positioned on the exterior of the coil. Most starches contain about 25% amylose (Table 6.1).

Starches also contain small amounts of an intermediate fraction with a degree of branching between that of amylose and that of amylopectin.

Two commercial high-amylose corn starches have apparent amylose contents of about 50 and 70%. Each contains the normal 20–25% amylose, but of low molecular weight (DP 250–300), and an amyloselike fraction that replaces approximately one-half and two-thirds, respectively, of the amylopectin molecules.

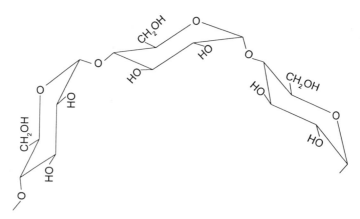

Fig. 6.3. Trisaccharide segment of unbranched portions of amylose and amylopectin molecules.

Linear segments of amylopectin molecules have the same right-handed α-helix conformation found in amylose molecules (Fig. 6.3). Pairs of nonreducing chain ends in clusters (Fig. 6.2) are entwined around each other in parallel double helices.

Granule Structure

As starch molecules form in an amyloplast, they nucleate, that is, they combine with one another, to form a compact, ordered mass that is semicrystalline. The region of ordered molecules continues to grow in a radial direction from the growth center, called the *hilum*. Completed granules, with molecules arranged in a radial direction, contain both polycrystalline and noncrystalline regions, in alternating layers. The clustered branches of amylopectin occur as packed double helices (Fig. 6.2). These double helical structures form the many small crystalline regions in the dense layers of starch granules that alternate with amorphous layers (Fig. 6.4). Because the crystallinity is produced by the ordering of the amylopectin chains, waxy starch granules have about the same amount of crystallinity as do normal starches. Amylose molecules occur among the

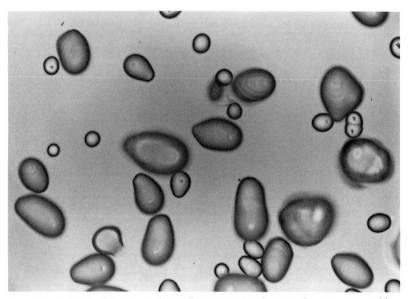

Fig. 6.4. A field of large and small potato starch granules as seen with a light microscope focused so that the layered arrangement of crystalline and amorphous regions can be seen in some of the large granules.

amylopectin molecules and can diffuse out of partially water-swollen granules.

The radial, ordered arrangement of starch molecules in a granule gives it a quasicrystalline nature, as evident from the polarization cross (birefringence) seen using a polarizing microscope (Fig. 6.5). The center of the cross is at the hilum. Cereal starches produce an x-ray pattern (type A) that is indicative of parallel, double helixes separated by interstitial water. In tuber and root starches, which produce B-type x-ray patterns, a column of water molecules replaces one of the double helices (Fig. 6.6). Amylopectin molecules forming the principal structure of the granule are arranged with their reducing ends toward the center of the granule.

Granule Types

Corn starch granules, even those from a single plant source, have diverse shapes (Fig. 6.7). Some in the floury center of the kernel endosperm are almost spherical. Others are more angular. Those in the proteinaceous vitreous (horny) endosperm found near the outside of the kernel are indented by protein bodies. Wheat starch granules

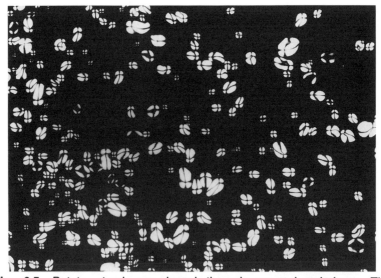

Fig. 6.5. Potato starch as viewed through crossed polarizers. The polarization cross (birefringence) is an indication of the ordered, quasicrystalline nature of native starch granules.

have a bimodal or trimodal size distribution due to initiation of granule growth at different times. In hard red winter wheat, A granules begin to form about two days after flowering and reach a

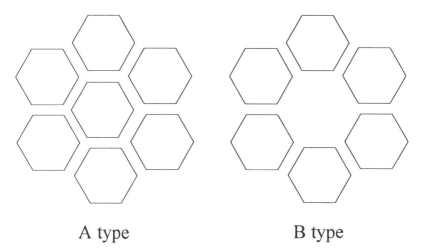

Fig. 6.6. Diagrammatic representation of the arrangement of six parallel double helices in starches that give a A-type pattern and starches that give a B-type pattern. Water molecules replace the center double helix in B-type starches.

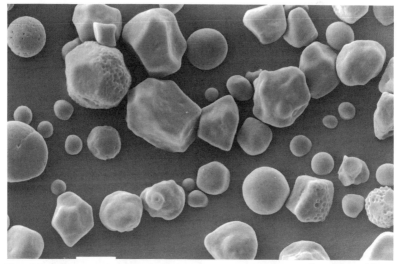

Fig. 6.7. Electron micrograph of common corn starch granules. Bar = 10 μm.

final diameter of at least 14 μm. B granules begin to form about 10 days after flowering and grow to a size of 5–16 μm. The larger wheat starch granules are lenticular. C granules, whose formation begins 21 days after flowering, grow to a size of only 1–5 μm. The approximate number yields of A, B, and C granules are 5, 50, and 46%, respectively; weight yields are 51, 45, and 3%, respectively. Soft wheat has two main granule types: A, large, ≈80% by weight, ≈50% by number; B, small, ≈20% by weight, ≈50% by number. Rice starch granules, on average, are the smallest of the commercial starches (2–9 μm), although small granules of wheat starch are almost the same size. Many of the granules in tuber and root starches, such as potato and tapioca starches, tend to be larger than those of seed starches and are generally less dense and easier to gelatinize. Some potato starch granules may be as large as 100 μm along the major axis. Potato and tapioca starch granules swell readily to large sizes. Banana starch, which can be obtained from green, cull bananas also has large granules. Amaranth starch granules are very small (about 1 μm in diameter).

Minor Components of Granules

All starches retain nonstarch components, including ash, lipid, and protein.

The phosphorus content of potato starch (0.06–0.1%) is due to the presence of phosphate ester groups. Calcium, magnesium, sodium, and potassium ions are also held in salt formation by the monoester phosphate groups, increasing the ash content. Cereal starches either do not have phosphate ester groups or have very much smaller amounts.

Only cereal starches contain significant endogenous lipids in the granules (Table 6.1). These internal lipids are primarily free fatty acids (FFA) and lysophospholipid (LPL), largely lysophosphatidylcholine (≈90% of the LPL in corn starch), with the ratio of FFA to LPL varying from one cereal starch to another. The lipids in common corn starch are 51–62% FFA and 24–46% LPL, while the lipids of wheat starch are 86–94% LPL and only 2–6% FFA. The lipids of barley starch are similar in composition to those of wheat starch. Almost equal proportions of LPL and FFA occur in millet, oat, rice, and sorghum starches. The lipid content of cereal starches varies with the amylose content. Nonwaxy cereal starches contain about 0.6–1.2% total lipid, which occurs both free and complexed with amylose. Waxy maize starch contains only about 0.2% lipid.

The amount of lipid also varies with the source. Root, tuber, and legume starches generally contain less lipid than do cereal starches. Potato starch contains only about 0.06% lipid, probably largely on the granule surface.

$$\begin{array}{c} \text{CH}_2\text{O}-\overset{\overset{\text{O}}{\|}}{\text{C}}-\text{R} \\ | \\ \text{HOCH} \\ | \\ \text{CH}_2\text{O}-\overset{\overset{\text{O}}{\|}}{\underset{\underset{\text{O}^-}{|}}{\text{P}}}-\text{O}-\text{CH}_2-\text{CH}_2\overset{+}{\text{N}}(\text{CH}_3)_3 \end{array}$$

FFAs and mono- or diacylglycerols, but not triacylglycerols, can form complexes with amylose by inclusion of one of the fatty acid chains within an amylose single helix (Fig. 6.8) (see Complexes). This interaction contributes to the low paste viscosity of cereal starches and to their gels being more opaque. Lipids complexed in helices are only extractable by polar solvents after pasting of the granules (see below).

Protein contamination of starch (corn, ≈0.35%; wheat, ≈0.4%) may be largely entrained endosperm material, but some protein is bound to the granule, at least in part as residual synthase enzyme. Inorganic material is present to an ash content of 0.1–0.5%.

Gelatinization and Pasting

Undamaged starch granules are not soluble in cold water but can imbibe water reversibly; that is, they can swell slightly and then return to their original size on drying. This reversible range varies. It

Fig. 6.8. Diagrammatic representation of a complex of a molecule of a fatty acid derivative with a segment of an amylose molecule.

is about 10% for normal corn starch and about 25% for waxy maize starch. Reversible swelling generally increases with granule diameter, approaching a doubling in size for the largest potato starch granules. The equilibrium moisture content of corn starch granules in water is about 28%, and of potato starch granules about 33%.

Gelatinization refers to the disruption of molecular order within starch granules as they are heated in the presence of water. Evidence for the loss of organized structure includes irreversible granule swelling, loss of birefringence (Fig. 6.5), and loss of crystallinity. Leaching of amylose occurs during gelatinization, but some leaching of amylose occurs at temperatures below the gelatinization temperature due to its location in noncrystalline regions and the fact that it is a relatively small, linear molecule that can diffuse out of granules. Gelatinization occurs over a temperature range, with larger granules generally gelatinizing first and smaller granules later. The apparent temperature of initial gelatinization and the range over which gelatinization occurs (see later) depend on the method of measurement and the starch-water ratio, granule type, and heterogeneities within the granule population. Values obtained using a polarizing microscope equipped with a hot stage are the initiation temperature (when the first granule in the field loses birefringence), the midpoint temperature, and the completion or birefringence endpoint temperature (when the last granule in the field loses birefringence). Other methods for determining the temperature and/or heat of gelatinization involve measuring the absorption of heat energy, loss of turbidity, molecule dissolution, dye adsorption, rate of enzyme-catalyzed hydrolysis, and chemical reactivity and determining changes in the x-ray pattern. Of these methods, one of the most sensitive, and one easy to measure, is the increase in extent of hydrolysis catalyzed by glucoamylase, or a mixture of α-amylase and glucoamylase, with the resulting D-glucose determined quantitatively using glucose oxidase (see section on Enzyme-Catalyzed Hydrolysis). Gelatinization temperature data for several commercial starches are given in Table 6.1.

Continued heating of starch granules in excess water results in further granule swelling and additional leaching of soluble components (primarily amylose). If shear is applied at this stage, granules are disrupted and a paste is formed. A starch paste is a viscous mass consisting of a continuous phase (a molecular dispersion) of solubilized amylose and/or amylopectin and a discontinuous phase of granule remnants (granule ghosts[3] and fragments). Complete molecular dispersion is obtained under conditions of high temperature and high shear and with excess water, conditions seldom, if ever, encountered in the preparation of food products. Cooling of a hot, corn-starch paste can result in a firm, viscoelastic gel.

Gelatinization and pasting of starch granules occurs because, as the temperature of a starch-water suspension increases, molecules within granules vibrate and twist so violently that intermolecular hydrogen bonds break and are replaced with hydrogen bonds to water molecules, producing more extensive hydration. Thus, starch polysaccharides become sheathed in layers of water molecules that plasticize them and allow them to move more freely. Eventually, molecular segments move to distances and positions that make it impossible for them to return to their original positions upon dehydration.

Gelatinization of starch is an energy-absorbing process that can be examined by differential scanning calorimetry (DSC), which measures both the temperature and the enthalpy of gelatinization (Fig. 6.9). Although there is not complete agreement on the interpretation of DSC data and the events that take place during gelatinization of starch granules, the following general picture is widely accepted. Water is a plasticizer for starch. Its mobility-enhancing effect is realized first in the amorphous regions, which are in a glassy, noncrystalline state.[4] When starch granules are heated in the presence of at least 60% water and their glass transition temperature, T_g, is reached, the plasticized amorphous regions of the granule undergo a phase transition from a glass to a rubber.[4] However, the energy absorption peak for this transition is not often seen because melting of the crystalline regions follows so closely upon the glass transition. Because the enthalpy of the initial melting endotherm (T_{m_1}) is large compared to that of the glass transition, the latter is usually not evident. Melting of lipid-amylose

---

[3] The ghost is the remnant remaining after cooking with shear or with only moderate shear. It is at least the remainder of the outer portion of the granule, the inner portion being solubilized. It has the appearance of an outer membrane, but starch granules have no membrane.

[4] A glass is a mechanical solid capable of supporting its own weight against flow. A rubber is an undercooled liquid that can exhibit viscous flow.

complexes, if present (see Complexes), occurs at much higher temperatures (100–120°C in excess water).

In normal food processing involving heat and moisture, starch granules quickly swell beyond the reversible point. Water molecules enter between chains, break interchain bonds, and establish hydration layers around the separating molecules. This effectively lubricates chains, which become more completely solvated and separated, causing tangential swelling of granules to several times their original size.

Cooking behaviors of different starches can be compared with a Brabender Visco/amylo/graph, which records the change in viscosity under low shear rates when the temperature is increased to 95°C, where it is held for a short time, then decreased (Fig. 6.10). When a 5% starch suspension is heated with gentle stirring, almost all water is absorbed within the granules, causing them to swell and press against each other, filling the container with a highly viscous starch paste. Highly swollen granules are fragile and easily broken by stirring, resulting in a decrease in viscosity. By the time peak viscosity is reached, some granules have ruptured and fragmented due to

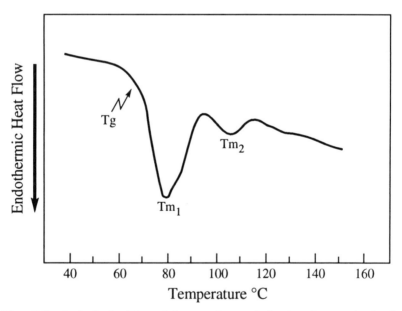

Fig. 6.9. A typical differential scanning calorimeter trace of starch gelatinization at 50% solids. T_g = glass transition temperature, T_{m_1} = initial melting endotherm, T_{m_2} = melting endotherm of amylose-lipid complexes.

shear forces. With continued stirring, more granules disintegrate, causing a greater decrease in viscosity. If the starch contains it, amylose will diffuse from swollen, gelatinized granules to the external water phase. Some amylopectin may also leach from them. A complete molecular dispersion is produced only from hot, dilute suspensions subjected to high shear, such as produced in specially designed cookers. Clarity of the suspension improves as granules swell and fragment.

On cooling, some starch molecules begin to reassociate, forming a precipitate or gel and an increase in paste opacity. This process is called *retrogradation* or *setback* (see Retrogradation and Staling). The firmness of the gel depends on the extent of junction zone formation (Chapter 5, Gels), which can be either facilitated or hindered by the presence of ingredients such as fats, proteins, sugars, and acids and the amount of water.

Tuber (potato) and root (tapioca) starches have weaker

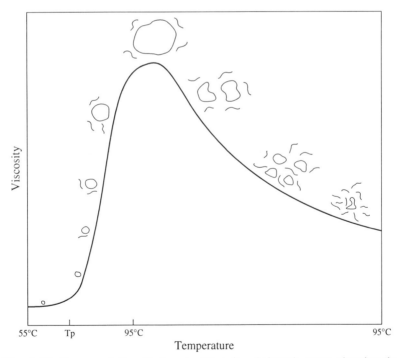

Fig. 6.10. Representative Brabender Visco/amylo/graph curve showing the typical swelling and disintegration that a granule undergoes during the cooking process in relation to viscosity.

intermolecular bonding and gelatinize easily to produce high-viscosity pastes that thin rapidly with moderate shear because their highly swollen granules break easily. Corn starch produces an opaque gel that undergoes syneresis (weeps or exudes moisture) and retains a slight cereal flavor. Waxy maize starch produces generally clear, noncohesive,[5] clean-flavored pastes. However, waxy maize starch can be easily overcooked to produce stringy pastes. Both corn starches are often modified before use in foods (see Starch Esters and Starch Ethers). High-amylose corn starches produce opaque, strong gels valuable for gum candies. Potato starch produces clear, viscous, almost bland pastes. It is subject to overcooking with loss of viscosity. Potato starch is used in extruded cereals and other such products and in dry mixes for soups and cakes. Tapioca starch gives clear, cohesive pastes that gel slowly and have a bland flavor. Rice starch produces opaque gels useful for baby food. Waxy rice starch gels are clear and cohesive. Wheat starch gels are weak and opaque and have a slight flavor.

Starches serve a variety of roles in foods and food production, principally to produce viscous pastes and soft-textured gels (Table 6.2). The extent of starch gelatinization in baked goods strongly affects product properties, including storage behavior and rate of digestion. In some baked products, many starch granules remain ungelatinized. For example, in certain cookies and pie crust, both high in fat and low in water, about 90% of wheat starch granules remain ungelatinized. In other products, such as angel food cake and white bread, which are high in moisture, about 96% of the wheat starch granules are gelatinized, although many become deformed.

During dry milling, some starch granules are damaged. Damaged starch absorbs more water and, when present in bread dough, produces a weak loaf and, often, a sticky crumb. Starch damage is also undesirable in soft wheat flour, which is used in cakes and cookies. Damaged granules are more susceptible to hydrolysis by α-amylase (see Enzyme-Catalyzed Hydrolysis), and this is the basis of one method for their detection.

The gelatinization temperature of starch granules is raised by the presence of neutral, water-soluble substances. Higher-molecular-weight carbohydrates are generally more effective than are mono-

---

[5] Cohesiveness is the tendency of the molecules to stick together. Cohesive gels are perceived as being difficult to break up in the mouth.

saccharides. However, the relationship is not based strictly on molecular weight, because sucrose raises the gelatinization temperature even more than a starch oligomer of DP 10 at an equal concentration. Sugars also decrease gel strength by interfering with the formation of junction zones between starch molecules.

Polar lipids retard starch gelatinization. They inhibit granule

TABLE 6.2
Typical Products Containing Native or Modified Starches

Bakery products	Dressings
Brownie mixes	Pourable salad dressings
Cake mixes	Spoonable salad dressings
Cream cakes	Sandwich spreads
Cupcake mixes	Frozen foods
Danish fillings	Frozen guacamole
Frostings	Frozen lasagna
Glazes	Frozen pot pies
Pie filling mixes	Frozen rice
Refrigerated doughs	Gravy mixes
Breadings for chicken patties,	Meat products
onion rings, fish products, etc.	Glazes
Beverages	Luncheon meats
Cocktail mixes	Poultry breast slices
Fruit juices	Sausage
Fruit drink mixes	Microwave meals
Canned foods	Preserves
Canned spaghetti	Rice noodles
Cream-style corn	Snacks, puffed
Condiments	Soups, etc.
Mustard	Chowders
Pickle relish	Dry soup mixes
Confectioneries	Soups
Candies	Stews
Caramel	Toppings
Fillings for chocolates	Butterscotch
Marshmallows	Marshmallow
Dairy products	
Cheesecakes	
Cheese powders	
Cheese spreads	
Dips	
Low-calorie cheesecakes	
Ice creams	
Sour creams	
Yogurts	

swelling primarily by complexing with amylose, and perhaps also by complexing with amylopectin (see Complexes).

Because starch is nonionic, salts have little effect on gelatinization and gel formation until their concentration reaches the point that they begin to lower water activity by solvation of large amounts of water, thus lowering the amount of water available for hydration of starch molecules, or by exerting an antiplasticizing effect. Such concentrations are usually not encountered in food products. Salts and sugars have less effect on potato amylopectin and ionic modified (phosphorylated) food starches (see Modified Food Starch). Molecules containing ionic groups have intensified water-binding ability at the ionic location, and, in addition, their repulsive charges cause the starch molecules to repel each other.

Water is an active ingredient of foods used to control texture and general physical behavior. It is not the total amount of water, but rather its availability, that is important. Starches and other carbohydrates absorb, hold, and control the migration of water.

Retrogradation and Staling

As already pointed out, cooling a hot starch paste generally produces a firm, viscoelastic gel, and gel formation is the result of junction zone formation (Chapter 5). The formation of junction zones can be considered to be the first stage of an attempt by starch (and other polymer) molecules to crystallize. Early starch users noted that, as starch pastes were cooled and stored, the starch became progressively less soluble. In dilute solution, starch molecules precipitated, and this insoluble material was difficult to redissolve by heating. They called this return to an insoluble state *retrogradation.* Retrogradation of cooked starch involves both of the two constituent polymers, amylose and amylopectin, with amylose undergoing retrogradation at a much more rapid rate than does amylopectin. The rate of retrogradation depends on several variables, including the molecular ratio of amylose to amylopectin; structures of the amylose and amylopectin molecules, which is determined by the botanical source of the starch; temperature; starch concentration; and presence and concentration of other ingredients, such as surfactants and salts. Many quality defects in food products, such as bread staling and loss of viscosity and precipitation in soups and sauces, are due, at least in part, to starch retrogradation.

Staling of baked goods is noted by an increase in crumb firmness and a loss in product freshness. Staling begins as soon as baking is complete and the product begins to cool. The rate of staling is dependent on the product formulation, the baking process, and storage conditions. Staling is due, at least in part, to the gradual transition of amorphous starch to a partially crystalline, retrograded state. In baked goods, where there is just enough moisture to swell and gelatinize starch granules (while retaining granule identity), and the degree of both swelling and gelatinization depends on the amount of water in the formulation, amylose retrogradation (insolubilization) may be largely complete by the time the product has cooled to room temperature. Retrogradation of amylopectin is believed to involve primarily association of its outer branches and requires a longer time than does retrogradation of amylose, giving it prominence in the staling process, which occurs over time after the product has cooled.

Amylose and amylopectin complexes are described in the next section. Most polar lipids with surfactant properties decrease crumb firming, their effectiveness in reducing retrogradation and staling being dependent on structure. Compounds such as glyceryl monopalmitate (GMP), other monoglycerides (especially glyceryl monomyristate and glyceryl monostearate) and derivatives of them, and sodium stearoyl 2-lactylate (SSL, a common name used in the industry) are incorporated into doughs of bread and other baked goods in part to increase shelf life.

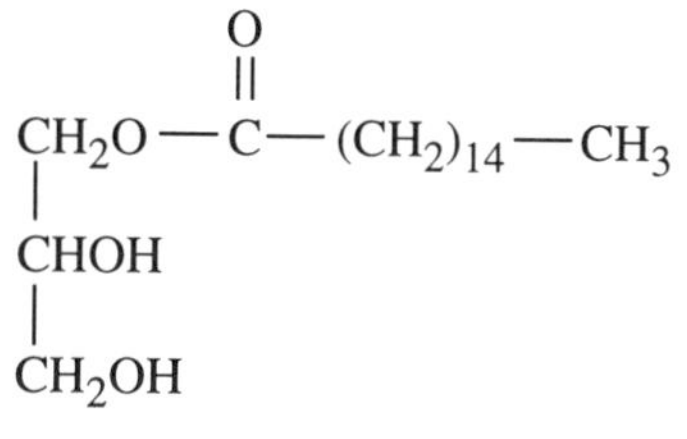

$$\begin{array}{l} CH_2O-\overset{\overset{\displaystyle O}{\|}}{C}-(CH_2)_{14}-CH_3 \\ | \\ CHOH \\ | \\ CH_2OH \end{array}$$

GMP

$$Na^+\,{}^-O_2C-\underset{\displaystyle CH_3}{\underset{|}{CHO}}-\left[\begin{array}{c}\overset{\overset{\displaystyle O}{\|}}{C} \\ | \\ CHO \\ | \\ CH_3\end{array}\right]_{0\text{-}4}-\overset{\overset{\displaystyle O}{\|}}{C}-(CH_2)_{16}-CH_3$$

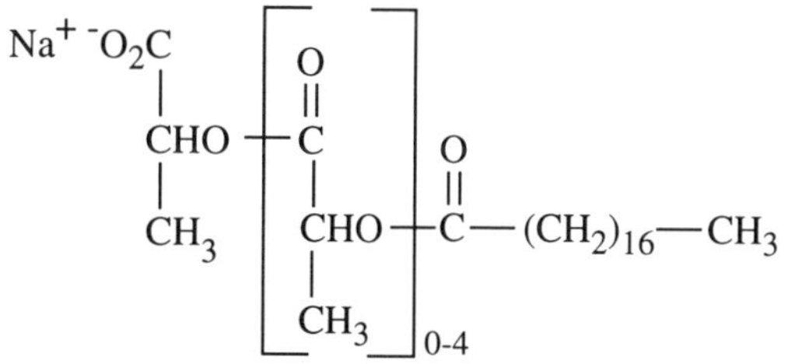

SSL

Complexes

Amylose chains are helical with hydrophobic, lipophilic interiors capable of forming complexes with linear hydrophobic portions of molecules that can fit within the lumen of the helix. When a hot, aqueous dispersion of starch is stirred with 1-butanol, nitropropane, or another polar hydrocarbon and then cooled, amylose complexes crystallize and can be isolated by centrifugation.

Iodine (as I_3^-) complexes with amylose and amylopectin molecules. Complexing occurs within the hydrophobic interior of helical segments. The long helical segments of amylose allow long chains of poly(I_3^-) to form, producing the blue color that is a diagnostic test for starch. The amylose-iodine complex contains 19% iodine, and determination of the amount of complexing is used to measure the amount of apparent amylose in a starch. Iodine colors amylopectin reddish purple because its branches are too short for formation of long chains of poly(I_3^-).

Polar lipids (surfactants or emulsifiers and fatty acids) affect starch pastes and starch-based foods in one or more of three ways as a result of complex formation: 1) by affecting starch gelatinization and pasting, 2) by modifying the rheological behavior of the resulting pastes, and 3) by inhibiting the crystallization of starch molecules associated with the retrogradation process. Specific changes observed upon addition of lipid depend on its structure, the starch employed, and the product to which it is added; but some generalizations can be stated. Different lipids/surfactants affect the gelatinization and pasting behaviors of a given starch differently, and each lipid/surfactant affects starches from different botanical sources differently. Addition of most lipids/surfactants to starches containing amylose inhibits the processes associated with gelatinization and pasting, but some cause the processes to be enhanced and/or to occur at a lower temperature. Polar native starch lipids inhibit gelatinization, pasting, and retrogradation. The helical complexes formed contain approximately six glucosyl units and six hydrocarbon methylene groups [-$(CH_2)_6$-] per helical turn (Fig. 6.8).

Polar lipids can complex at almost any point over the entire length of amylose molecules. Up to 86% of the glucosyl units may be complexed in amylose molecules saturated with a lipid. Amylopectin can bind only about 15% as much lipid per unit weight, complexation probably being restricted to the outer chains.

The ability of polar lipids to form complexes with amylose and amylopectin is associated with their chain length and degree of unsaturation and with the nature of their hydrophilic group. In general, esters of myristic (C_{14}, saturated) and palmitic (C_{16}, saturated) acids are most effective.

Complexes between starch molecules (particularly amylose) and fatty acyl compounds are digested by α-amylase very slowly.

Hydrolysis

Starch molecules, like other polysaccharide molecules, are depolymerized by hot acids, the reaction becoming faster as the temperature is increased. Hydrolysis occurs more or less randomly and initially produces linear and branched fragments. Commercially, hydrochloric acid is sprayed onto well-mixed starch or stirred moist starch is treated with hydrogen chloride gas, and the mixture is heated until the desired degree of hydrolysis is obtained. The acid is then neutralized, and the product is recovered by filtration or centrifugation, washed, and dried. This process is called *thinning*. The products, called *acid-modified* or *thin-boiling* starches, are still granular, but break up (cook out) easily and dissolve when heated in water. Even though only a few glycosidic bonds are hydrolyzed, the starch granules disintegrate much more easily during heating in water. Acid-modified starches form gels with improved clarity and increased strength but provide less viscosity. Thin-boiling starches are used as film formers and adhesives in products such as pan-coated nuts and candies and for situations where strong gels are desired, for example, in gum candies such as jelly beans, jujubes, orange slices, and spearmint leaves and in processed cheese loaves. To prepare especially strong and fast-setting gels, high-amylose corn starch is used.

More extensive modification with acid produces *dextrins*. Low-viscosity dextrins can be used in relatively high concentrations. They have film-forming and adhesive properties somewhat similar to those of natural gums. They are used as coating agents in products such as pan-coated roasted nuts and candy, fillers, encapsulating agents, and carriers of flavors, especially spray-dried flavors. They are classified by their cold water solubility and color, but mainly by their *dextrose equivalency* (DE). The DE is related to the degree of polymerization through the following equation: $DE = 100 \div DP$. (Both DE and DP are average values for a population of molecules.) Thus, the DE of a hydrolysis product is its reducing power as a percentage of the reducing power of pure dextrose (D-glucose), and

DE is inversely proportional to average molecular weight. Products with a low DE value retain large amounts of linear chains or long chain fragments, allowing them to form strong gels.

Hydrolysis of hot starch dispersions with either an acid or an enzyme(s) to DE values of less than 20 produces *maltodextrins*. Maltodextrins made by acid catalysis have a high quantity of linear chains and are able to retrograde, forming a haze. To prevent haze and to produce maltodextrins of low hygroscopicity and high water solubility, acid-catalyzed hydrolysis to DE 5–10 is followed by enzyme-catalyzed hydrolysis with a *Bacillus* α-amylase (see next section). Maltodextrins are also produced by use of an α-amylase alone and with a combination of an α-amylase and a debranching enzyme. Maltodextrins of lowest DE are nonhygroscopic, while those of high DE, that is, low average molecule weight, absorb moisture. Maltodextrins are bland, with virtually no sweetness, and are excellent for contributing bulk to food systems. In general, maltodextrins are used in dry powder mixes, coffee whiteners, snack foods, bakery products, confections, imitation cheeses, frozen foods, and sauces because of their easy dispersibility, rapid solubility, film-forming ability, low hygroscopicity, and blandness, and for their bulking, binding, and cryoprotectant properties.

Hydrolysis to DE 20–60 gives mixtures of molecules that, when dried, are called *corn syrup solids*. They dissolve rapidly and are mildly sweet.

Continued hydrolysis of starch with an acid and/or enzymes produces a syrupy mixture of D-glucose, maltose, other malto-oligosaccharides, and a few products that result from new glycosidic bond formation at the high concentrations used in commerce. The latter process is called *reversion* (Chapter 1). Such syrups are produced in enormous quantities (see D-Glucose and D-Fructose Production). The most common has a DE of 42. These starch hydrolysates are stable because crystallization does not occur easily in such complex mixtures. They are sold in concentrations of sufficient osmolality to prevent growth of ordinary microorganisms. In the United States, such syrups are called *corn syrups*. In Europe, they are known as *glucose syrups*. An example is waffle and pancake syrup, which is a corn/glucose syrup that has been colored with caramel coloring and flavored with maple flavoring.

Uses of maltodextrins and corn (glucose) syrups in foods is widespread. They are used largely as humectants and to provide bulk and body and have degrees of sweetness that can vary from little or none to a sweetness greater than that of sucrose or invert sugar (Chapter 3) (see D-Glucose and D-Fructose Production). Products with the greatest sweetness (solutions of D-glucose and D-fructose) are used to prepare still and carbonated beverages and a variety of processed foods, including bakery products, dairy products, pickles, jams, jellies, preserves, canned fruits, confections, and meat products. Those with moderate sweetness are used in foods requiring body, but not high sweetness, such as sauces, toppings, and some confections and sweet bakery products. Products with low sweetness provide body to ice cream, canned fruits and vegetables, bakery products, and some confections. Both maltodextrins and corn (glucose) syrups inhibit crystallization of sucrose and water and are useful in confections and frozen foods and desserts for this reason.

Hydrolyzing Enzymes

Three to four enzymes are used for industrial hydrolysis of starch to D-glucose. α-Amylase and glucoamylase are the two that attack undamaged starch granules. Under nonswelling conditions, both slowly increase the porosity of corn starch granules, apparently breaking down molecules in the less dense regions.

α-Amylase

Α-Amylase is an endoenzyme that cleaves both amylose and amylopectin molecules internally. α-Amylases from different sources have different action patterns and produce different product mixtures. One that is commonly used industrially produces large amounts of maltopentaose (G5)[6] through maltononaose (G9). A typical digest after long incubation will contain G6–G12 > G5≫G3≫G2>G4>G1. The larger oligosaccharides may be singly, doubly, or triply branched via (1→6) linkages, since α-amylase acts only on the (1→4) linkages of starch. α-Amylase does not attack double-helical starch polymer segments or polymer segments complexed with a polar lipid (stabilized single helical segments).

Glucoamylase

Glucoamylase,[7] in combination with an α-amylase, is the enzyme used commercially for producing D-glucose (dextrose) syrups and

---

[6] G5, for example, is a shorthand designation for a maltooligosaccharide containing five α-D-glucopyranosyl units.

[7] Also known as amyloglucosidase.

crystalline D-glucose. It acts upon fully gelatinized starch as an exo-enzyme, sequentially releasing single D-glucosyl units from the nonreducing ends of amylose and amylopectin molecules, even those joined through (1→6) bonds. Such bonds become available as amylopectin molecules are eroded down to branch points. Consequently, the enzyme can completely hydrolyze starch to D-glucose. Commercial end products, however, contain only about 95% D-glucose because the reaction in concentrated solutions reaches an equilibrium with a small reversal producing about 1% each of maltose, isomaltose, and other disaccharides and 2–3% of higher oligosaccharides.

β-Amylase

β-Amylase releases maltose sequentially from the nonreducing ends of amylose and amylopectin. It cannot cleave the (1→6) linkages at branch points. Thus, in the case of amylopectin, it leaves a pruned molecule termed a *limit dextrin*, specifically a β-*limit dextrin*. Since there is not always an even number of D-glucopyranosyl units in the outer branches of amylopectin, there will be two or three D-glucopyranosyl units left attached as a branch (A chain) and one or two D-glucopyranosyl units left in the main chain (B chain) beyond the branch. To produce maltose, starch is hydrolyzed with hydrochloric acid to a DE of about 20; then the digest is neutralized and hydrolyzed with malt β-amylase.

Debranching Enzymes

There are several debranching enzymes that specifically catalyze hydrolysis of (1→6)-linkages in amylopectin, producing numerous linear but lower-molecular-weight molecules that still produce an intense blue color with iodine. One such enzyme is called *isoamylase*; another is *pullulanase*.

Cyclodextrin Glucanotransferase

Cyclodextrin glucanotransferase is a unique *Bacillus* enzyme that forms rings of (1→4)-linked α-D-glucopyranosyl units from amylose and amylopectin. The enzyme can form six-, seven-, and eight-membered rings (Fig. 6.11). The normal helical conformation of a linear portion of a starch molecule contains six to seven glucosyl units per turn of the helix. Transfer of a glycosidic bond from one that joins adjacent units of the spiral to one that forms a doughnut-like circular structure produces products, originally called "Schardinger dextrins" after their discoverer, but now known as *cyclodextrins* or *cycloamyloses*. The six-, seven-, and eight-membered rings are respectively, α-, β-, and γ-cyclodextrins. They have the ability to complex with hydrophobic substances that are held in the center of the ring. Such complexes are termed *clathrates*, and the molecules complexed are known as *guest molecules*. This complexation is useful for fixing aromas and flavors. In this way, volatile essential oils can be converted into dry powders in which the flavoring or aromatic substance is protected from light and oxygen but is readily released when the complex is added to an aqueous system because of the water solubility of the cyclodextrin. Cyclodextrins are not approved for food use in the United States. Chiral supports that are useful for chromatographic separations are made by converting cyclodextrins into insoluble polymeric materials. Insoluble polymeric beads of cyclodextrins have also been shown to be useful for removal of the bitter components of citrus juices.

D-Glucose and D-Fructose Production

Corn (glucose) syrup is the major source of D-glucose and D-fructose. A glucose syrup is made by first passing a slurry of starch in water containing a thermally stable α-amylase through a jet cooker, where rapid gelatinization, complete granule disruption (due to the high shear), and enzyme-catalyzed hydrolysis (liquefaction) occur. After the solution is cooled to 55–60°C (130–140°F), glucoamylase is added and hydrolysis is continued. When hydrolysis is complete, the syrup is clarified, decolorized, and concentrated.

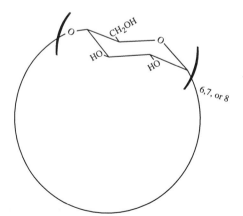

Fig. 6.11. Structures of the three cyclodextrin molecules.

When crystalline D-glucose (dextrose) or its monohydrate are desired, seed crystals are added.

For production of D-fructose, a solution of D-glucose is passed through a column containing bound glucose isomerase, which catalyzes isomerization of D-glucose to D-fructose to an equilibrium mixture of approximately 58% D-glucose and 42% D-fructose. Higher concentrations of D-fructose are usually desired. To make the high-fructose corn syrup used as a soft drink sweetener (approximately 55% D-fructose), the isomerized syrup is passed through a bed of cation-exchange resin in the calcium salt form to bind D-fructose, which is then removed to provide an enriched syrup fraction. Crystalline D-fructose is available commercially.

Modified Food Starch

Food processors generally prefer starches with better behavioral characteristics than provided by native starches. Corn (maize) and waxy corn (waxy maize) starches produce weak-bodied, cohesive, rubbery pastes and undesirable gels when cooked, but their functional properties can be improved by modification. Modification increases the ability of starch pastes to withstand the heat, shear, and acid associated with processing conditions and introduces specific functionalities. Modified food starches are functional and abundant food macroingredients and additives.

Types of modifications that are made, singly or in combinations, are crosslinking of polymer chains, noncrosslinking derivatization (sometimes called stabilization), depolymerization, and pregelatinization. Specific property improvements that can be obtained are reduction in the energy required for cooking, modification of cooking characteristics, increased solubility, increased or decreased paste viscosity, increased paste stability, increased freeze-thaw stability of pastes, enhancement of paste clarity, increased paste sheen, inhibition of gel formation, enhancement of gel formation and gel strength, reduction of gel syneresis, improvement of interaction with other substances, improvement in stabilizing properties, enhancement of film formation, improvement in water resistance of films, reduction in paste cohesiveness, and improvement of stability to acid, heat, and shear (Table 6.3).

The most important commercial derivatives of starch are those in which only a very few of the hydroxyl groups are reacted. Normally ester or ether groups are attached at very low degree of substitution

(DS) values (Chapter 5). DS values are usually <0.1 and generally in the range 0.002–0.2. Thus, on average, there is one substituent group on every 500 to five D-glucopyranosyl units. Derivatized granules appear the same as underivatized granules, but the small levels of derivatization dramatically change the properties of starches and greatly extend their usefulness and broaden their applications. Inserted groups, such as hydroxypropyl ether groups, restrict intermolecular association and the ability of starch chains to form junction zones. Thus, gels of these derivatized starches remain stable and do not undergo syneresis.[8] Such starch products that are esterified or etherified with monofunctional reagents and resist interchain associations are called *stabilized starches*. Use of difunctional reagents produces *crosslinked starches*.

Starches are most often modified in an aqueous slurry. For esterification or etherification, a starch slurry of 30–45% solids from wet milling (in the case of corn starches) is introduced into a stirred reaction tank. Sodium sulfate or sodium chloride is added to 10–30% concentration to inhibit gelatinization. The pH is adjusted,

---

[8] Syneresis is the expulsion of water from interstices of a gel caused by the sliding of polymer chains over each other to form longer lengths of hydrogen-bonded chains, thereby constricting space for water in the gel's three dimensional network (Chapters 4 and 5).

TABLE 6.3
Some Characteristics Imparted to Starches By Modification

Modification	Major Characteristics
Pregelatinization	Solubility and dispersion without cooking. Cold water swelling.
Esterification	
Acetylation	Improved paste clarity and stability at low temperatures. Easier cooking.
Phosphorylation	Improved paste clarity and stability. Freeze-thaw stability.
Octenylsuccinylation	Emulsifying and emulsion stabilizing properties.
Phosphate ester cross-linking	Increased viscosity and body. Stability to heat, shear, and acidic conditions. Storage stability. Delayed pasting.
Etherification/hyrdoxpropylation	Improved paste clarity and stability. Freeze-thaw stability. Easier cooking.

generally with sodium hydroxide, to pH 8–12, the exact value depending on the reaction. The chemical reagent is added. Temperature is generally controlled to less than 60°C (140°F) to prevent pasting so that the derivatized starch can be recovered in granule form. Following reaction to the desired DS, the starch is recovered by filtration or centrifugation, washed, and dried.

Chemical reactions currently allowed and used to produce modified food starches in the United States are as follows:
1. Esterification with acetic anhydride, succinic anhydride, the mixed anhydride of acetic and adipic acids, 1-octenylsuccinic anhydride, phosphoryl chloride, sodium trimetaphosphate, sodium tripolyphosphate, and monosodium orthophosphate
2. Etherification with propylene oxide
3. Acid modification with hydrochloric and sulfuric acids
4. Bleaching with hydrogen peroxide, peracetic acid, potassium permanganate, and sodium hypochlorite
5. Oxidation with sodium hypochlorite

and various combinations of these reactions. Modified food starches are often both crosslinked and stabilized.

Stabilized Starches

Derivatization of starches with monofunctional reagents reduces intermolecular associations, gelation, development of opacity, and precipitation. Hence, such derivatization is called *stabilization*. Pastes of unmodified common corn starch produce opaque, cohesive,[9] rubbery, long-textured, and syneresing gels. Waxy maize starch pastes are relatively clear and have little tendency to gel at room temperature, making waxy maize starch the preferred base starch for most modified food starches. However, waxy maize starch pastes become cloudy and chunky and exhibit syneresis when stored under refrigerator or freezing conditions. The most common derivatives employed for starch stabilization are the acetate ester, the monostarch phosphate ester, and the hydroxypropyl ether.

Starch Esters

Acetylation of starch to the maximum allowed in foods (DS 0.09) lowers the gelatinization temperature, improves paste clarity, and provides stability to retrogradation and freeze-thaw cycling (but not

---

[9] See footnote 5 for an explanation of cohesiveness.

as well as hydroxypropylation; see Starch Ethers). Upon cooking, a higher peak viscosity is obtained. Upon cooling of the resulting paste, the viscosity is lower than that obtained from the unmodified starch, an indication of improved stability.

Starch phosphate monoesters made by drying starch in the presence of sodium tripolyphosphate or monosodium orthophosphate produce clear, stable pastes that have freeze-thaw stability. Monostarch phosphates have a long, cohesive texture. Paste viscosity is generally high and can be controlled by varying the concentration of reagent, time of reaction, temperature, and pH. Salt reduces the viscosity. Phosphate esterification lowers the gelatinization temperature. The maximum DS allowed in the United States is 0.002. Very little monostarch phosphate is used in foods.

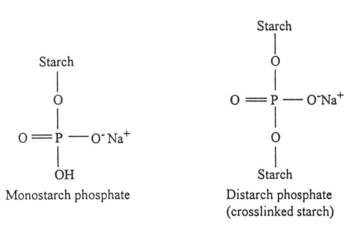

Monostarch phosphate

Distarch phosphate
(crosslinked starch)

Preparation of an alkenylsuccinate ester is a way to attach a hydrocarbon chain to starch. The hydrocarbon chain is first attached to a dibasic acid anhydride; then the anhydride is reacted with starch to make a monoester. The product is low-DS starch 1-octenylsuccinate. It disperses well because the bulky substituent group keeps molecules separated. It is compatible with fat or oil because of the hydrophobicity of the alkenyl group. Hence, even at very low DS, starch 1-octenylsuccinate molecules concentrate at the interface of an oil-in-water dispersion, making the product an emulsifier. Starch 1-octenylsuccinate is also an emulsion stabilizer because it is polymeric. It can be used in a variety of food applications where emulsion stability is needed, such as in pourable dressings and flavored beverages. The presence of the aliphatic chain tends to give the starch derivative a sensory perception of fattiness, so it is possible to use the derivative as a partial replacement for fat in certain foods.

$$\text{Starch—OH} + \text{CH}_3\text{—(CH}_2)_5\text{—CH=CH—HC} \underset{\underset{\displaystyle \text{H}_2\text{C—C}}{|}}{\overset{\overset{\displaystyle \text{O}}{\|}}{\text{C}}} \underset{\|}{\overset{\diagdown}{\text{O}}} \\ \downarrow \\ \text{Starch—O—}\overset{\overset{\displaystyle \text{O}}{\|}}{\text{C}}\text{—CH—CH=CH—(CH}_2)_5\text{—CH}_3 \\ \phantom{\text{Starch—O—C—}}|\\ \phantom{\text{Starch—O—C—}}\text{CH}_2 \\ \phantom{\text{Starch—O—C—}}| \\ \phantom{\text{Starch—O—C—}}\text{CO}_2^-$$

Crosslinked Starches

Most modified food starches are crosslinked. Crosslinking occurs when granules are reacted with difunctional reagents that react with hydroxyl groups on different molecules. Most crosslinking is accomplished by producing distarch phosphate esters. Phosphate ester crosslinking occurs when starch is reacted with phosphoryl chloride, $POCl_3$, or sodium trimetaphosphate in an alkaline slurry. The linking together of starch chains with phosphate diester or other crosslinks strengthens the granule and reduces both the rate and the degree of granule swelling and subsequent disintegration. Thus, crosslinked starch granules are less sensitive to processing conditions (high temperature; extended cooking times; low pH; high shear during mixing, milling, homogenization, and/or pumping) than are native granules. Cooked pastes of crosslinked starches are more viscous, heavier bodied, shorter textured, and less likely to break down during extended cooking, exposure to low pH conditions, or when subjected to severe agitation. Only a small amount of crosslinking is required to produce an effect. With low levels of crosslinking, granule swelling is inversely proportional to DS. Thus, treatment of starch with only 0.0025% of sodium trimetaphosphate[10] reduces both the rate and the degree of granule swelling, greatly

---

[10] If the reaction were 100% efficient, this amount of reagent would provide a degree of substitution of about 1.3×10^{-5} or one crosslink for 77,000 D-glucopyranosyl units.

increases paste stability, and dramatically changes the Brabender Visco/amylo/graph viscosity profile (Fig. 6.12) and textural characteristics of its paste. Treatment with 0.08% trimetaphosphate produces granules with such restricted swelling that peak viscosity is never reached in a Brabender Visco/amylo/graph. As the number of crosslinks increases, the granules become more and more stable to physical conditions and acidity, but less and less dispersible by cooking. Energy requirements to reach maximum swelling and viscosity are also increased. Even though hydrolysis of glycosidic bonds occurs during heating under acidic conditions, chains tied to each other through phosphate crosslinks continue to provide large molecules and elevated viscosity in low-pH systems.

While other crosslinking agents exist, the only other crosslink permitted in a food starch is the distarch ester of adipic acid.

$$\text{Starch} - O - \overset{\overset{O}{\|}}{C} - CH_2 - CH_2 - CH_2 - CH_2 - \overset{\overset{O}{\|}}{C} - O - \text{Starch}$$

Most crosslinked food starches contain less than one crosslink per 1,000 α-D-glucopyranosyl units. Crosslinking of waxy maize starch

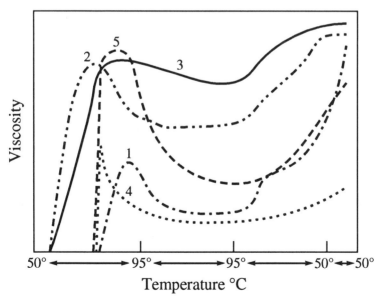

Fig. 6.12. Typical Brabender Visco/amylo/graph curves: 1 = common corn starch, 2 = stabilized common corn starch, 3 = moderately crosslinked and stabilized common corn starch, 4 = waxy maize starch, 5 = potato starch.

gives a clear paste that, when used as a pie filling, for example, is sufficiently rigid that the filling does not boil out when heated.

Crosslinked starches that form stable, high-viscosity pastes when dispersions are subjected to high shear, high temperature, and/or low pH (Fig. 6.12) are particularly useful in continuous processes. High levels of crosslinking are useful for extruded products. Storage-stable thickening is provided by crosslinked starches. Such starches are useful in retort sterilization of canned foods because of their reduced rate of gelatinization and swelling. This allows the low initial viscosity to remain for sufficient time to facilitate the heat transfer and temperature rise required to provide uniform sterilization before granule swelling develops the desired viscosity, texture, and other properties of the final product. Crosslinked starches are used in spoonable salad dressings, canned soups, canned gravies, canned puddings, and batter mixes.

Crosslinked, stabilized starches are used in canned, frozen, baked, and dry foods. They provide stability to products such as baby foods and fruit pie fillings in cans and jars, frozen fruit pies, pot pies, and gravies during long-term storage.

Starch Ethers

Hydroxypropylstarch is prepared by reacting starch with propylene oxide to produce low levels of etherification (DS 0.02–0.2, 0.2 being the maximum allowed). Its properties are similar to those of starch acetate because the appended groups also produce "bumps" along the polymer chains. Thus, it is a stabilized starch, has a lowered gelatinization temperature, and produces clear pastes that do not retrograde and that withstand freezing and thawing. Hydroxypropylstarches are used as thickeners and extenders. Lower-molecular-weight products are good coffee whiteners. To improve viscosity under acidic conditions, acetylated and hydroxypropylated starches are often also crosslinked with phosphate groups.

Cold-Water-Soluble (Pregelatinized) Starch

Once starch has been pasted and dried without excessive molecular reassociation, it can be dissolved in water at room temperature. The major amount of such starch is made by flowing a starch-water slurry onto a steam-heated roll or into the nip between two nearly touching and counter-rotating, steam-heated rolls, where the starch is quickly gelatinized, pasted, and dried. The dry film is scraped from the roll and ground. The resulting product should contain no intact granules, except in the case of a crosslinked starch. Such starches, known as *pregelatinized* or *instant* starches, are precooked starches. They are also prepared using extruders.

Pregelatinized starches can be used without cooking, producing dispersions without lumps if coarsely ground. Finely ground pregelatinized starch behaves similarly to a water-soluble gum, forming small gel particles (see Chapters 4 and 5) and producing some graininess or pulpiness, which is desirable in some products. Many pregelatinized starches are used in dry mixes such as instant pudding mixes; they disperse readily with high-shear stirring or when mixed with sugar or other dry ingredients. Both chemically modified and unmodified starches can be used to make pregelatinized starches. If chemically modified starches are used, properties introduced by modification(s) are also exhibited by the pregelatinized products; thus, paste properties, such as stability to acid, shear, and freeze-thaw cycles, are also characteristics of pregelatinized starches. For example, pregelatinized, slightly crosslinked starch (see Starch Esters) yields a paste of high shear strength useful in instant soup, pizza topping, and extruded snacks. It also has application in limited-moisture systems such as in soft cookies and bakery fillings.

Starches that are not pregelatinized are known as *cookup* starches.

Cold-Water-Swelling Starch

A starch that swells extensively in water at room temperature is made by heating common corn starch in 75–90% ethanol to 150–175°C (300–345°F) for 0.5–2.0 hr. This product is also pregelatinized. The product described in the previous section, while historically called pregelatinized starch, is actually prepasted starch. A similar cold-water-swelling starch can also be made by very quickly heating a starch slurry in a special spray-drying nozzle. Cold-water-swelling starch is dispersible in sugar or corn syrups by rapid stirring, as in a Waring Blendor. The resulting dispersion can be poured into molds, where it sets to a rigid gel that can be sliced easily to make a gum candy. The ability to swell in unheated water is also useful for making desserts and in muffin batters that contain particles, such as blueberries, that would settle to the bottom when the batter thins by heating (before baking, that is, before the thickening effect of the wheat starch is realized). Treating corn starch in either way causes disruption of many intermolecular hydrogen bonds. When cold-water-swelling granules are placed in water, hydration and swelling are rapid. In this condition, retrogradation reintroduces insolubility and poor quality of aqueous dispersions. However, when the treated granules are dispersed in a sugar solution, retrogradation is minimal.

A further modification of the process of heating starch in aqueous ethanol converts a mixture of corn and waxy maize starch to an agglomerate that readily produces high viscosity when added to water. Waxy maize starch alone, on heating in aqueous ethanol, losses its granule structure and, on addition to water, dissolves to produce, on cooling, a hard candylike product.

Multiple Modifications and Imparted Functionalities

Modified food starches are tailor-made for specific applications. Many modified food starches are made by a combination of crosslinking and introduction of monosubstituent groups (stabilization). Acid-thinning and/or pregelatinization may also be employed.

Food product functionalities that can be controlled by selection of the proper modified starch include, but are not limited to, the following: adhesion, clarity, color, emulsion stability, film formation, flavor, flavor release, hydration rate, moisture retention and control, mouthfeel, oil migration control, texture/consistency, physical state (liquid, semisolid, solid), sheen, shelf stability, acid stability, heat stability, shear stability, tackiness, temperature required to cook, and hot and cold viscosity.

Manufacture of Starch

Corn is the major source of starch because of its abundance and low price. Grain is cleaned of metal and dust, placed in steep tanks with water and sulfur dioxide at pH 3–4, and held at 48–52°C (120–125°F) for 30–40 hr. Softened grain is then shredded in a mill that does not crush the germ. Germ, consisting of oil and protein, is removed in continuous liquid cyclones,[11] and the remainder of the

---

[11]Liquid cyclones or hydroclones are cone-shaped tubes with two outlets, one at the top and one at the bottom, and an inlet port located on the side near the top. A suspension is forced in through the inlet in a way that produces a rotational velocity (a vortex) in the fluid. The centrifugal force produced by the rotational velocity is sufficient to separate particles based on their density, with the heavier particles passing out the bottom of the tube and the lighter ones out the top.

grain is ground more finely to free starch granules. After screening out of the seed coat particles, the starch slurry is concentrated in smaller hydroclones, then either introduced into a modification tank or passed to centrifuges or filters, where the starch is recovered, then dried.

Potato tubers and tapioca cassava roots do not require steeping; potatoes, for example, are put directly through a hammer mill with sulfur dioxide water. The starch is separated from the fiber by use of appropriate screens. Thereafter, the starch is removed in cyclones, washed, and dried.

Chapter 7

Cellulosics

Cellulose is the principal cell-wall component of higher plants and hence the most abundant organic compound, and the most abundant carbohydrate, on Earth.[1] It is a high-molecular-weight, linear, insoluble homopolymer of repeating β-D-glucopyranosyl units joined by (1→4) glycosidic linkages. Because of their linearity and stereoregular nature, cellulose molecules associate over extended regions, forming polycrystalline, fibrous[2] bundles, in which the molecular chains in crystalline regions are held together by numerous hydrogen bonds. Cellulose is insoluble except in a few special solvents that can disrupt these intermolecular bonds. However, certain derivatives of cellulose are water soluble and important as food gums (see Carboxymethylcellulose, Methylcelluloses, and Hydroxypropylmethylcelluloses).

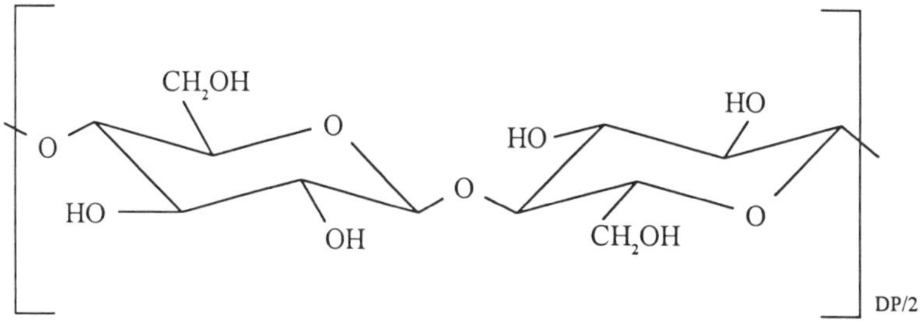

Cellulose

---

[1] When cellulose is considered as a polymer of D-glucose, D-glucose can be said to be the most abundant carbohydrate and even the most abundant organic compound on Earth.

[2] The term *fibrous* indicates that the lengths of the bundles of molecules are far greater than their widths.

Characteristic properties of cellulose derivatives are shown in Table 7.1.

Cellulose

Cellulose and its modified forms serve as dietary fiber (Chapter 15) because no forms of cellulose are digested by humans and thus none provide significant nourishment or calories. However, dietary fiber does serve important functions (Chapter 15).

TABLE 7.1
Characteristic Properties of Cellulosics

Product	Major Characteristics
Cellulose powders	Provide noncaloric bulk
	Retain moisture
Microcrystalline celluloses	Stabilize foams
	Stabilize emulsions
	Replace fats and oils
	Form thixotropic gels
	Improve adhesion (cling)
	Provide freeze-thaw stability
	Modify texture
	Retard ice crystal growth
	Anticaking activity in grated and shredded cheese
Carboxymethylcelluloses	Thicken
	Retard ice crystal growth
	Form films
	Bind and hold water
	Protect colloids
	Processing acid
	Humectant
	Retard sugar crystallization
	Prevent syneresis
	Stabilize proteins
Methylcelluloses and hydroxypropylmethylcelluloses	Thermal gelation
	Reduce amount of fat required
	Provide lubricity
	Form and stabilize emulsions
	Form and stabilize foams
	Form films
	Provide freeze-thaw stability

The commercial source of cellulose is wood pulp or cotton linters, the short fibers remaining on cotton seeds after the long fibers are removed. Because cotton fibers are about 98% cellulose, cotton linters need only a treatment with a hot sodium hydroxide solution that removes protein, pectic substances, and wax to produce high-quality cellulose. Wood (about 50% cellulose, 30% hemicellulose [Chapter 8], and 20% lignin) is pulped to solubilize the latter two components. In the pulping (delignification) operation, wood chips are digested with calcium bisulfite in the presence of sulfur dioxide (the bisulfite process), with alkaline sodium sulfide (the sulfate process), or with sodium hydroxide (the soda process). The latter process is the primary source of cellulose used to make water-soluble derivatives for the food industry. This pulp is further purified by treatment with alkali and an alkaline solution of sodium hypochlorite to remove additional amounts of hemicellulose and color and the remaining traces of lignin.

Powdered Cellulose

A purified cellulose powder obtained by special pulping of wood is available as a food ingredient. A measure of the quality of cellulose is its content of α-cellulose, which is that portion insoluble in 18% alkali. β-Cellulose is the portion of pulp that dissolves in 18% alkali but precipitates when the solution is neutralized. γ-Cellulose is the fraction remaining soluble after neutralization of the 18% alkali solution. Highly purified pulp with an α-cellulose content of >99% can be considered to be pure (1→4)-linked β-glucan. While the more purified forms are used to make the cellulose gums (sodium carboxymethylcellulose, methylcellulose, and hydroxypropylmethylcellulose, described later), chemical purity is not required for food use because cellulosic cell-wall materials are components of all fruits, vegetables, flours, meals, and brans. The powdered cellulose used in foods has negligible flavor, color, and microbial contamination. It is available in fibers varying in length from 0.5 to 4 mm and in width from 0.005 to 0.35 mm. To make powdered cellulose, pulp that has been produced by delignification of wood chips is purified and bleached until it meets the specifications of the *Food Chemicals Codex*.

Powdered cellulose may be added to bread to provide noncaloric bulk. Reduced-calorie baked goods containing powdered cellulose have an increased content of dietary fiber (Chapter 15) and stay moist and fresh longer. In frozen novelties, such as ice pops, powdered cellulose maintains texture through freeze-thaw cycles. In sauces, it provides a smooth, creamy texture and increases cling and viscosity. Powdered cellulose of at least 0.11-mm particle size increases the viscosities of solutions of guar gum (Chapter 9), sodium carboxymethylcellulose (this chapter), and xanthan (Chapter 10) because of the interactions of these gums with cellulose molecules.

Other products in this category are microfibrillated and microreticulated celluloses. Bacterial cellulose gel (from *Acetobacter xylinum*), called *nata*, is a favorite dessert delicacy in the Philippines.

Microcrystalline Cellulose

Microcrystalline cellulose (MCC), is made by hydrolysis of purified wood pulp (α-cellulose pulp), followed by separation of the cellulose microcrystals. Cellulose molecules are fairly rigid, completely linear chains of about 3,000 (1→4)-linked β-D-glucoyranosyl units that align easily in long junction zones, the first step in crystallization. However, the long and unwieldy chains do not fit together over their entire lengths. So the crystallites have no regular surface. The end of the crystalline region is simply the divergence of cellulose chains away from order into a more random, amorphous, spaghetti-like arrangement. When purified wood pulp is treated with acid, hydronium ions most easily penetrate the lower-density, noncrystalline regions and cause hydrolytic cleavage of the chains, thereby releasing individual, fringed crystallites. The released crystallites grow larger because the chains that constitute the fringes now have greater freedom of motion and can order themselves (Fig. 7.1).

Two general types of microcrystalline cellulose are produced. One is powdered MCC, which is a spray-dried product. Spray-drying produces agglomerated aggregates of microcrystals that are porous and spongelike. Average particle sizes in different products generally range from about 20 to about 90 μm. A particle 30 μm in diameter typically contains 6×10^8 microcrystals (0.1 μm in diameter). Powdered MCC is used primarily as an anticaking agent and flavor carrier for grated and shredded cheese. It serves to stabilize dispersed oil, producing a semisolid. It is also used to make reduced-calorie and/or high-fiber bakery products and as an extrusion aid for expanded snacks and restructured, frozen, french-fried potatoes.

The second type of MCC is colloidal MCC. It is water dispersible and has functional properties somewhat similar to those of water-soluble gums. To make colloidal MCC, considerable mechanical energy is applied after hydrolysis to tear apart the weakened microfibrils and provide a major proportion of colloidal-sized aggregates (<0.2 μm in diameter). To prevent rebonding of the aggregates during drying, sodium carboxymethylcellulose (CMC) (see Carboxymethylcellulose) is added. CMC aids in redispersion and acts as a barrier to reassociation by giving the particles a stabilizing negative charge from the ionized carboxyl groups of the adhering polymer molecules. The CMC also aids in redispersion of the microcrystals and leashes them together after dispersion (Fig. 7.2).

MCC does not swell or dissolve in water. Viscosity development is dependent on shear to disperse the particles. During high shear in water over a 10-min period, particles progress from being partially hydrated to being fully swollen, and the partial dispersion of microcrystals becomes a fully dispersed colloidal suspension of microcrystals. The viscosity of the dispersion continues to increase over 12–24 hr as the dispersed microcrystals hydrate and swell. This increase can be 10-fold, for example, from 100 to 1,000 mPa·sec. The nature of the suspension varies with various factors (including the particular MCC product used, the temperature, and other ingredients), but generally, as used, it forms a thixotropic gel (Chapter 5). The rheology and stability of the system can be modified through the use of other gums that act as colloid protectors. Xanthan works

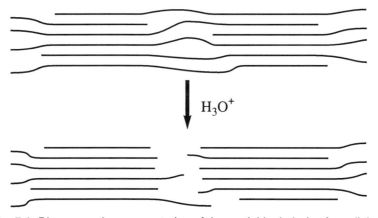

Fig. 7.1. Diagrammatic representation of the partial hydrolysis of a cellulose fiber to produce the microcrystals of microcrystalline cellulose. Because the chains in the amorphous regions have greater freedom of motion after cleavage, the crystals can grow in size and are actually larger in the product than they were in the original pulp.

best as a protective colloid; CMC and hydroxypropylmethylcellulose (see Methylcellulose) are also effective. Xanthan is the only gum that is effective in very acidic systems, for example, pH 3.

Both types of MCC are relatively stable to heat and acids. Their major functions are 1) to stabilize foams and emulsions, even during high-temperature processing; 2) to form thixotropic gels with salvelike textures, although they do not dissolve, but rather form networks of hydrated microcrystals as opposed to intermolecular junction zones (Chapter 5); 3) to stabilize pectin and starch gels to heating; 4) to modify textures; 5) to improve adhesion; 6) to replace fat and oil (Chapter 15); and 7) to control ice crystal growth (provide freeze-thaw stability/heat shock improvement). MCC stabilizes emulsions and foams through its property of collecting at oil-water and air-water interfaces and strengthening interfacial films.

Colloidal-type MCM is used in ice cream at concentrations of 0.2–0.4%. It disperses as small particles, some less than 0.2 μm in diameter. In ice cream, it acts as a suspending agent, enhances stiffness, and retards ice crystal growth. It is a common ingredient in reduced-fat ice cream and ice-creamlike products. In salad dressings, it improves uniformity, stability, pasteurizability, and spreadability. It is not an emulsifier but tends to collect at interfaces of oil-in-water emulsions, where it acts to thicken the mixture and to improve mouthfeel. It improves the cling of sauces, even at elevated

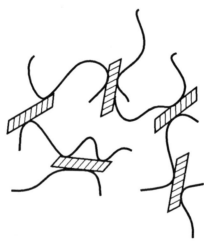

Fig. 7.2. Diagrammatic representation of the tying together of cellulose microcrystals with sodium carboxymethylcellulose (not drawn to scale).

temperatures, and improves the quality of low-solids tomato sauce. With starch, it improves thickening with less masking of flavor. It significantly improves the body, texture, and stability of a 20–27% vegetable fat whipped topping. Typical products in which MCC and other cellulosic derivatives are used are shown in Table 7.2.

Coprocessed mixes of MCC and guar gum (Chapter 9) and MCC and an alginate are particle aggregates that can mimic the rheology of fat-in-water emulsions and contribute fatlike mouthfeel to nonfat and low-fat products (Chapter 15).

Ingredient labels on food products containing MCC may read only *microcrystalline cellulose* or *cellulose gel*.

TABLE 7.2
Typical Products Containing Microcrystalline Cellulose, Carboxymethylcellulose, Methylcellulose, and Hydroxypropylmethylcellulose

Microcrystalline Celluloses	
Barbecue sauces	Low-calorie spoonable salad dressings
Frozen cheese lasagna	Marshmallow topping
Frozen guacamole	No-oil pourable salad dressings
Fruit drink mixes	Sandwich spreads
Grated cheese	Shredded cheese
Liquid diet products	Whipped toppings
Low-calorie mayonnaise	

Carboxymethylcelluloses	
Cake mixes	Ice creams
Cheesecakes	Instant cappuccinos
Cupcake mixes	Low-calorie cheesecakes
Diet dairy drink mixes	Low-calorie mayonnaise
Dips	Marshmallow topping
Frozen chicken sticks	Pancake syrups
Frozen turkey nuggets	Pasta salad mix
Fruit drink mixes	Shredded cheese
Grated cheese	Snack cakes
Hot cocoa mixes	Sour cream

Methylcelluloses	
Bakery products	Fruit juices
Barbecue sauces	Onion rings
Frozen novelties	Sandwich spreads
Fruit juice mixes	Twice-baked potatoes

Hydroxypropylmethylcelluloses	
Cream cheese	Low-calorie cheesecakes
Honey-roasted peanuts	Sandwich spreads

Carboxymethylcellulose

CMC, the sodium salt of the carboxymethyl ether of cellulose (cellulose-O-CH$_2$-CO$_2^-$Na$^+$) is widely used as a food gum. Treatment of purified wood pulp with 18% sodium hydroxide solution produces "alkali cellulose." In alkali cellulose, many of the hydroxyl groups have been converted to alkoxide groups that can undergo Williamson ether synthesis by reaction with an alkyl halide. For production of CMC, alkali cellulose is reacted with the sodium salt of chloroacetic acid. In the following equation, $m + n = 3$ and $x + y = 3$ (because there are three hydroxyl groups per glucosyl unit) and $y =$ the degree of substitution (DS).

$$\text{Cell-(OH)}_m(\text{O}^-\text{Na}^+)_n + \text{ClCH}_2\text{CO}_2^-\text{Na}^+$$
$$\downarrow$$
$$\text{Cell-(OH)}_x(\text{O-CH}_2\text{CO}_2^-\text{Na}^+)_y + \text{NaCl}$$

Carboxymethylcellulose

Commercial sodium CMC has a DS (see Chapter 4) of less than 1.5, but most products have DS values in the range 0.4–0.8. Products most often used as food ingredients have a DS of 0.7. For water solubility, CMC must have a DS of 0.4 or greater. Substituent groups are found at O-2, O-3, and O-6 of D-glucopyranosyl units in the proportion 2.14:1.00:1.58. The degree of uniformity of derivatization along the cellulose chain determines solution behavior. Nonuniform distribution leaves unsubstituted stretches of molecules that are available for junction zone formation with unsubstituted regions of other CMC molecules, producing thixotropy (see Chapter 5) and binding to cellulosic products. CMC binds to MCC and interacts with certain other water-soluble gums such as locust bean gum (Chapter 9). Uniform derivatization produces molecules without unsubstituted stretches that cannot bind to other CMC molecules, thereby producing smooth, stable solutions. CMC is available in a wide range of viscosity types. Solution viscosity depends mainly on the average molecular weight of the gum (Table 7.3).

Since CMC consists of long, fairly stiff molecules bearing negative charges, its molecules in solution are stretched out due to electrostatic repulsion of chain segments. In addition, chains repel each other, producing monodisperse, highly viscous, stable solutions. However, lowering the pH to <4 represses ionization of carboxyl groups so that some lose their charge. Molecular association then occurs, and viscosity increases. At pH 3, insolubilization occurs due to extensive charge removal. Under acidic conditions, CMC undergoes hydrolysis with loss of viscosity, especially when its solutions are heated.

$$\text{Cell-(OH)}_x(\text{O-CH}_2\text{CO}_2^-\text{Na}^+)_y$$
$$\downarrow H_3O^+$$
$$\text{Cell-(OH)}_x(\text{O-CH}_2\text{CO}_2^-\text{Na}^+)_{y-z}(\text{O-CH}_2\text{CO}_2\text{H})_z$$

CMC stabilizes protein dispersions, especially near their isoelectric pH value. Thus, milk products are stabilized against casein precipitation, as CMC forms stable soluble complexes with casein at pH values in the range 3–6 where casein by itself is insoluble. Other typical applications of CMC are shown in Table 7.4.

Ingredient labels on products containing this gum may read *sodium carboxymethylcellulose, sodium carboxymethyl cellulose, carboxymethylcellulose, carboxymethyl cellulose, CMC, sodium CMC,* or *cellulose gum.*

Methylcelluloses and Hydroxypropylmethylcelluloses

Alkali cellulose can be treated with methyl chloride to introduce methyl ether groups. In this reaction, the protons of some hydroxyl groups are replaced by methyl groups (cellulose-O-CH$_3$), creating methycellulose (MC). Most members of this class of food gums also contain hydroxypropyl ether groups (cellulose-O-CH$_2$-CHOH-CH$_3$). Hydroxypropylmethylcelluloses (HPMCs) are made by reacting

Table 7.3
Viscosity Types of Carboxymethylcellulose

Viscosity	Average Degree of Polymerization	Average Molecular Weight
High	3,200	700,000
Medium	1,100	250,000
Low	400	90,000

TABLE 7.4
Typical Applications of Carboxymethylcellulose

Product Type	Functions
Cake and related mixes	Batter thickener
	Humectant (improves texture and extends shelf life)
Cheese spreads	Protective colloid
Dietetic foods	Thickener
	Bodying agent
Dressings	Thickener
Dry pet food	Makes gravy when water is added
Dry-powder fruit drink mixes	Suspending aid
Extruded products	Lubricant
	Binder
	Film former
	Processing aid
Fillings	Holds moisture and prevents syneresis
Frozen and dried egg white	Protein stabilizer
Hot cocoa mixes	Thickener
Ice cream and other frozen dessert products	Retards ice crystal growth
	Improves mouthfeel, body, and texture
Icings and frostings	Retards sugar crystallization
Meat emulsions	Texturizer
	Binder
Milk products	Protein stabilizer
Puddings	Holds moisture and prevents syneresis
Sauces	Suspending aid
Syrups	Thickener
Toppings	Holds moisture and prevents syneresis

alkali cellulose with both propylene oxide and methyl chloride. Both the MC and HPMC members of this gum family are generally referred to simply as *methylcelluloses*. They are cold-water soluble because the protrusions along the chains prevent intermolecular association. They are not readily hydrolyzed by cellulase enzymes and hence are somewhat resistant to microbial attack. The methyl ether DS (Chapter 4) levels of commercial MCs range from 1.1 to 2.2. The hydroxypropyl molar substitution (Chapter 4) levels in commercial HPMCs are in the range 0.02–0.3. At this low degree of reaction, it is unlikely that much formation of poly(propylene oxide), i.e., poly(propylene glycol), chains occurs.

$$\text{Cell-(OH)}_m(\text{O}^-\text{Na}^+)_n + \text{ClCH}_3$$
$$\downarrow$$
$$\text{Cell-(OH)}_x(\text{O-CH}_3)_y + \text{NaCl}$$
Methylcellulose

$$\text{Cell-(OH)}_m(\text{O}^-\text{Na}^+)_n + \text{ClCH}_3 + \overset{\displaystyle\diagup\text{O}\diagdown}{\text{CH}_2-\text{CH}-\text{CH}_3}$$
$$\downarrow$$
$$\text{Cell-(OH)}_x(\text{O-CH}_3)_y(\text{O-CH}_2\text{-CHOH-CH}_3)_z + \text{NaCl}$$
Hydroxypropylmethylcellulose

While a few ether groups spread along cellulose molecules enhance water solubility by prevention of chain association, they also decrease chain hydration by replacing water-binding hydroxyl groups with less polar ether groups. Thus, the ether groups, while responsible for water solubility by retarding interchain association, restrict hydration of the chains to the point that they are on the borderline of water solubility. Hence, when an aqueous solution is heated, the water molecules solvating the polymer molecules are given sufficient kinetic energy that many dissociate from the chain, allowing intermolecular associations and gelation to occur. Reducing the temperature once again brings about rehydration and solubility. Thus, the thermal gelation is reversible. This property is the basis for many of the applications of MCs.

MCs can also be used to reduce the amount of fat in food products through two mechanisms: 1) they impart fatlike properties so that the fat content of a product can be reduced and 2) they reduce absorption of fat by products being fried. (It is believe that the gel structure produced by thermal gelation provides a barrier to fat and oil and holds moisture. Thermal gelation also causes MCs to function as binders.) They impart richness (creaminess), provide lubricity (slippery mouthfeel), provide structure and body, and generate and stabilize foams. In baked products, they retain moisture and thereby improve tenderness and extend shelf life, and thermal gelation gives greater gas retention during baking. When foods to be fried, such as doughnuts and onion rings, are formulated with an MC in the batter, the result is a reduction in fat absorption, improved adhesion, and an increase in moisture retention, because the gel structure produced by thermal gelation provides a barrier to oil, holds moisture, and acts as a binder. Low-molecular-weight MCs interact synergistically with hydroxypropylstarch (Chapter 6).

The major use of HPMC products is in nondairy whipped toppings. At about 0.5% concentration, HPMC stabilizes foams, imparts better whipping characteristics, prevents phase separation through its surface activity, and provides freeze-thaw stability by becoming more soluble as the temperature is lowered.

Acceptable label designations for MC are *methylcellulose* and *methyl cellulose*; *modified vegetable gum* is used occasionally. Acceptable label designations for HPMC are *hydroxypropyl methylcellulose, hydroxypropylmethylcellulose*, and *hydroxypropyl methyl cellulose*; *carbohydrate gum* is used occasionally.

Chapter 8

Hemicelluloses

Hemicelluloses are plant cell-wall polysaccharides. They vary in amount and in structure, depending on the plant type and the location of the cells within the plant. Generally they constitute 20–30% of cell walls, but in annual food plants, they are present in even greater amounts. They may be roughly defined as those land plant cell-wall polysaccharides other than cellulose and pectin. The designation *hemi*, meaning *half*, was included as part of the name by early investigators because of their belief that the hemicelluloses were on their way to becoming cellulose. However, hemicelluloses constitute a separate family of structurally unrelated polysaccharides. They are components of both the soluble and insoluble fractions of the dietary fiber of all plant materials, especially brans, vegetables, and fruits (see Chapter 15). In addition, they give fruits and vegetables crisp, chewy, fibrous characteristics that partially remain on cooking.

Structures and Characteristics

Many hemicelluloses are fairly long polysaccharide chains with frequent branches consisting of one to several sugar units, often of different sugar types. The most common hemicelluloses in annual plants, especially in farm crops, consist of a core composed of a chain or chains of (1→4)-linked β-D-xylopyranosyl units. Some are essentially linear molecules with only a few short side chains. Others are highly branched, bushlike structures with short side chains on a central core structure. The short side chains are from one to a few units long and are usually composed of L-arabinofuranosyl, D-galactopyranosyl, D-glucuronopyranosyl, and/or 4-*O*-methyl-D-glucuronopyranosyl units. Occasionally present are L-rhamnopyranosyl, L-arabinopyranosyl, L-galactopyranosyl, and/or L-fucopyranosyl units and/or various methylated sugars (Fig. 4.1). Often these polysaccharides occur as partially acetylated polymers, with acetyl contents of up to 12% by weight.

Most hemicelluloses can be isolated from plant tissue by extraction with alkaline solutions. They can be separated by solubility. Those polysaccharides classified as hemicellulose A precipitate upon neutralization of the alkaline extract and are essentially linear polymers. A second group of polysaccharides, the hemicellulose B fraction, precipitates on addition of ethanol to the neutralized extract to a concentration of 60–70%. Hemicellulose B components are more highly branched, more-soluble polymers.

In food preparation and cooking, some soluble hemicelluloses may dissolve and be lost. Those remaining contribute to the physical properties of the system. Rich sources of hemicelluloses are seed coats from commercially processed grains such as corn, wheat, oats, barley, and rice. Other possible sources are skins from sugar beets, potatoes, and tomatoes. These are all available at low cost, being byproducts or waste residues.

β-Glucan

A water-soluble polysaccharide present in cereal brans[1] and known as β-glucan has become a commercial product because of its effectiveness in reducing serum cholesterol (see also Chapter 15). Oat and barley brans are incorporated into a variety of foods and are the source of commercial β-glucan. The principal material used for β-glucan extraction is the crude oat or barley bran fraction that contains portions of starchy endosperm and has a β-glucan content of at least 5.5%, usually 6.6%. β-Glucan is extracted with water at 65–100°C (150–212°F), and the extract is dried on heated rolls or by spray-drying. Extraction with sodium carbonate solution at pH 10 can be used at lower temperatures.

Oat β-glucan is a linear chain of β-D-glucopyranosyl units; about 70% are linked (1→4) and about 30% (1→3). The (1→3) linkages occur singly and are separated by sequences of, generally, two or three (1→4) linkages. Thus, the molecule is composed of (1→3)-linked β-cellotriosyl and cellotetraosyl units (Fig. 8.1), i.e.,

---

[1] *Bran* is the term used to designate the outer layers of cereal grains.

$$\longrightarrow 3)-\beta Glcp-(1 \longrightarrow \left[4)-\beta Glcp-(1 \longrightarrow \right]_n$$

Fig. 8.1. Representative structure of a segment of oat and barley β-glucans where *n* usually is 1 or 2 but occasionally may be larger.

$$\rightarrow 3)\text{-}\beta Glcp\text{-}(1\rightarrow 4)\text{-}\beta Glcp\text{-}(1\rightarrow 4)\text{-}\beta Glcp\text{-}(1\rightarrow .$$

Such (1→4,1→3)-β-glucans are often called *mixed linkage β-glucans*.

Nutritionally, oat and barley β-glucans are classified as "soluble dietary fiber" (Chapter 15). When taken orally in foods, they reduce postprandial serum glucose levels and the insulin response in normal and diabetic human subjects, perhaps due to increased viscosity of the intestinal contents, in addition to reducing serum cholesterol concentrations. These physiological effects are typical of those of soluble dietary fiber. Other soluble polysaccharides have similar effects to differing degrees.

A bacterial β-glucanase is used in the preparation of wort for the brewing process. The enzyme reduces the viscosity of barley, rice, rye, and other cereal grain malt and flour dispersions by hydrolyzing the β-glucans into tri- and tetrasaccharides, thereby reducing wort viscosity, increasing the rate of wort filtration, and improving beer clarity and stability.

Corn Fiber Gum

Corn fiber, a fraction containing fragments of the seed coat produced as a by-product in the wet milling of corn, is the source of a hemicellulose. It can be obtained by lime water (dilute calcium hydroxide solution) extraction of corn fiber, neutralization of the extract with carbon dioxide, removal of calcium ions as insoluble calcium carbonate, and precipitation of the hemicellulose from the concentrated solution with isopropanol or by roll-drying.

Corn fiber gum is completely water soluble and has a bland flavor and uniform composition. It has properties suggesting usefulness in foods as a low-viscosity thickener and/or an emulsion-stabilizing and -extending agent. An emulsion of corn oil in water is stabilized by corn fiber gum.

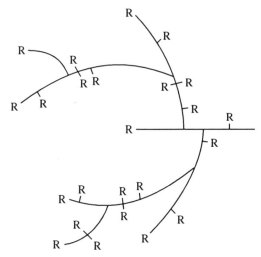

Fig. 8.2. Representative structure of corn fiber gum. Longer lines represent the xylan backbone structure. R = GlcpA, LAraf, LArap, D and LGalp, LGalp-Xylp-LAra, Xylp-LAra, LAra-LAra, LAra-Xylp-LAra, and perhaps other units. (See Fig. 4.1.)

The gum consists of a highly branched, substituted chain of (1→4)-linked β-D-xylopyranosyl units; that is, it is a highly branched xylan. About 23% of the xylosyl units are branched, the branch points being (1→3) linkages. Various side chains, largely L-arabinofuranosyl units, are attached to the main chain. Some oligosaccharide chains of two or three sugar units are also present (Fig. 8.2). It is a relatively low-molecular-weight polysaccharide.

Wheat bran hemicellulose is similar to corn fiber gum in structure and composition.

Larch Arabinogalactan

Another low-viscosity, water-soluble gum can be extracted with hot water from a number of softwoods, but particularly from chips of the western larch, *Larix occidentalis*. Larch arabinogalactan was once approved as a food additive, but its approval was dropped for lack of use. It consists of a backbone chain of (1→3)-linked β-D-galactopyranosyl units, each of which bears a substituent at the O-6 position. Most of the side-chain substituents are (1→6)-linked β-galactobiosyl units. Other common side chains are single L-arabinofuranosyl and 3-O-(β-L-arabinopyranosyl)-α-L-arabinofuran-

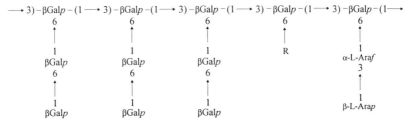

Fig. 8.3. Representative structure of larch arabinogalactan. R = a β-D-galactopyranosyl unit or, less frequently, an α-L-arabinofuranosyl or a β-D-glucopyranosyluronic acid unit. The location of the side chains is not intended to indicate order or frequency, only typical disorders and the type of structure that could be expected in a segment of the gum.

osyl units (Fig. 8.3). Arabinogalactans have been recommended for use as emulsifiers, stabilizers, and binders in flavor bases, dressings, and pudding mixes.

Chapter 9

Guar and Locust Bean Gums

Sources and Structures

Guar and locust bean gums are important thickening polysaccharides for both food and nonfood uses. Guar gum produces the highest viscosity among the natural, commercial gums, with a 1% solution having a viscosity of 6,000–10,000 mPa·sec.

Both gums are the ground endosperm of seeds. Guar plants look somewhat like soybean plants; they have an erect stem with seeds in numerous pods in clusters, hence the sometime-used name *cluster bean*. A galactomannan is the main component of both endosperms. Galactomannans have a main chain of $(1\rightarrow4)$-linked β-D-mannopyranosyl units, bearing single α-D-galactopyranosyl units attached to O-6 of about 56% of the main-chain units (Fig. 9.1).

The polysaccharide of guar gum is guaran. Commercial guar gum, the ground endosperm of guar seeds, usually contains 80–85% guaran, 10–14% moisture, 3–5% protein, 1–2% fiber, 0.5–1.0% ash, and 0.4–1.0% lipid. It is available as powders of various grades and mesh sizes.

Commercial locust bean gum (LBG, also called carob gum) is the ground endosperm of locust bean (carob) seeds. It, too, is available as powders of various grades. Locust bean seeds are obtained from the large pods of locust, or carob, trees. These trees grow around the Mediterranean Sea, where they were planted by early traders who used the uniformly sized seeds for measurement of weight. The weight terminology has carried over to the present measurement of diamonds in *karats*; originally one karat was the weight of one carob seed. Pods dropped in the fall by locust trees are collected and taken to kibblers, who separate the seeds for milling and grind the empty pod to a powder that may be used for extending chocolate.

Locust bean galactomannan has fewer side chains (branches) than does guaran and its structure is irregular. It has long stretches of about

80 underivatized D-mannopyranosyl units alternating with sections of about 50 units in which almost every main chain unit has an α-D-galactopyranosyl group glycosidically connected to its O-6 position.

Properties

Because of the differences in polysaccharide structures, guar gum and LBG have quite different physical properties, even though both galactomannans are composed of long, rather rigid chains that have a large hydrodynamic volume, thus providing high solution viscosity. Guar gum, especially, produces high viscosity with pseudoplastic rheology at low concentrations.

Because guaran has its galactosyl units fairly evenly placed along the chain, there are few locations on the chains that are suitable for significant junction zone formation. However, LBG, with its long "naked chain" sections, can form junction zones and produce solutions with thixotropy. Furthermore, LBG molecules can interact with bare portions of cellulose derivatives (Chapter 7) to form junctions and to produce a synergistic[1] increase in viscosity. LBG also interacts with xanthan (Chapter 10) and carrageenan (Chapter 11) helices; in these cases, rigid gels are formed (Fig. 9.2).

Guar gum hydrates and builds viscosity rapidly when dispersed in water. Although different mesh sizes hydrate and produce viscosity at different rates, generally about half of the final viscosity is obtained in 10–15 min without heating; about 1 hr is required for full viscosity

---

[1] A synergistic interaction is one in which the observed value is greater than what would be predicted by summing the parts.

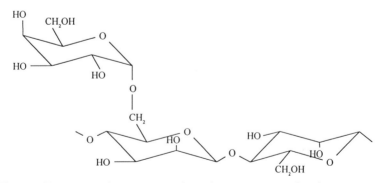

Fig. 9.1. Representative segment of a galactomannan molecule.

development. A higher viscosity is obtained after first heating then cooling a dispersion of guar gum, an effect that is much more pronounced with LBG.

Guar and locust bean gums, because they are neutral molecules, are compatible with most other food substances, including other polymers. Guar gum does, however, exhibit interactions with starches (Chapter 6), cellulose (Chapter 7), agar, κ-carrageenan (Chapter 11), and xanthan (Chapter 10). The interaction results in binding in the case of cellulose and a synergistic increase in viscosity in the case of water-soluble polysaccharides. For example, when a combination of guar gum and xanthan is used, the observed viscosity is greater than would be predicted if the molecules of guar gum and the molecules of xanthan acted independently, indicating an intermolecular interaction (Fig. 9.3). The xanthan-guar gum ratio required for maximum viscosity varies with ionic strength from about 50:50 to about 90:10 but is usually about 70:30. The intermolecular interaction of guar gum with cellulose is used to make a fat mimic (Chapter 7).

Because guar gum consists of long, soluble, rather rigid molecules, it is able to reduce fluid friction during pumping of aqueous liquids by suppressing turbulent flow. Guar gum solutions tolerate salt well, but high concentrations of calcium salts produce precipitation, as do many polyvalent cationic salts.

Because LBG molecules contain "naked" regions—pure β-(1→4)-

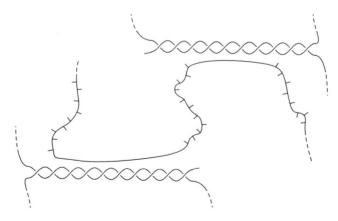

Fig. 9.2. Representation of the hypothesized interaction of a locust bean gum molecule with helical portions of carrageenans and/or xanthan to form a three-dimensional structure and a gel.

mannan segments—that fit together in much the same way as do cellulose, i.e., β-(1→4)-glucan, molecules, LBG is only slightly soluble in room temperature water; most particles swell only partially. Heating a suspension to about 85°C (185°F) is required for good dissolution. Cold-water-soluble modifications are available. While LBG solutions do not form significant gels alone, they do when mixed with agar, κ-carrageenan (Chapter 11), or xanthan (Chapter 10). For maximum interaction between LBG and the other gums, the hydrated mixtures need to be heated to temperatures at least above 60°C (140°F) to at least partially dissolve the LBG. With κ-carrageenan, for example, an increased gel strength (over that obtained with κ-carrageenan alone) is realized. In addition, the texture of the gel is modified. LBG makes carrageenan gels less brittle and more elastic by replacing carrageenan-carrageenan junctions with weaker mixed-polymer junctions. Syneresis is also reduced. When a hot solution containing xanthan (a nongelling polymer) and LBG (another nongelling polymer) is cooled, a gel is formed because of junction zone formation between the two polymers (Figs. 9.2. and 9.4).

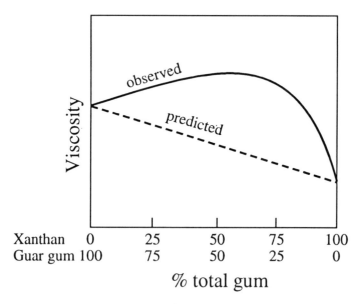

Fig. 9.3. Graph indicating the synergistic interaction between guar gum and xanthan to give viscosities greater than that predicted by summing the two viscosities, indicating that they do not act independently of each other.

Uses

Guar Gum

Guar gum provides economical thickening to numerous food products. Its main applications are in dairy products, prepared meals, bakery products, sauces, and pet food. It is frequently used in combination with other food gums, particularly in dairy products. For example, in ice cream and related products, where it is used as a stabilizer, CMC (Chapter 7), carrageenan (Chapter 11), xanthan (Chapter 10), and LBG are also common ingredients. The prime function of guar gum is to bind water; it also prevents ice crystal growth, improves mouthfeel, reduces the chewiness produced by a combination of carrageenan and LBG, and slows meltdown (see Chapter 11).

Guar gum minimizes syneresis in processed cheese. Significant amounts of guar gum are used in canned and intermediate-moisture pet foods because of its high viscosity, heat stability, and low cost.

It improves mixing and recipe tolerance in bakery products such as specialty breads, cakes, and doughnuts and also improves shelf life

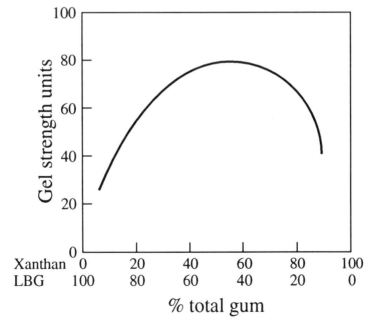

Fig. 9.4. Gel strengths of the gels formed by mixtures of locust bean gum (LBG) and xanthan, each by themselves nongelling gums.

through moisture retention. It prevents syneresis in frozen foods and pie fillings and controls spreadability in prepared icings.

Locust Bean Gum

About 85% of LBG is used in dairy and frozen dessert products. It is rarely used alone but in combination with other gums such as CMC, carrageenan, xanthan, and guar gum (see Chapter 11). In ice cream and related products, it provides excellent heat shock resistance, smooth meltdown, and desirable texture and chewiness. A typical use level is 0.05–0.25%.

In soft cheese, such as cream cheese, LBG speeds coagulation, increases curd yield, and makes curd separation easier. It enhances resilience and body and prevents syneresis in the finished cheese. It pro-

TABLE 9.1
Typical Products Containing Galactomannans

Guar Gum

Condiments	Egg substitutes
Horseradish sauce	Frozen foods
Pickle relish	Fish sticks
Salsa	Frozen cheese products
Dairy products	Frozen novelties
Cheese spreads	Frozen onion rings
Cold pack cheese	Frozen turkey nuggets
Cottage cheese	Frozen twice-baked potatoes
Cream cheese	Instant hot cereals
Dips	Low-calorie salad dressings
Frozen cheese products	Mixes
Frozen novelties	Cake mixes
Frozen yogurt	Dry soup mixes
Ice creams	Pasta salad mix
Low-calorie cheesecake	Pet foods
Low-fat yogurt	Sloppy Joe sauce
Processed cheese	Spanish olive pimento stuffing
Sour cream	

Locust Bean Gum

Bakery products	Dips
Dairy products	Frozen novelties
Cheese spreads	Ice creams
Cottage cheese	Whipped toppings
Cream cheese	Low-calorie salad dressings

vides excellent spreadability and stability to cheese spreads and may be used in imitation sour cream, sour cream-based dips, and yogurt.

LBG can serve as a binder and stabilizing agent in processed comminuted meats, such as sausages, to improve homogeneity and texture. It has a lubricating effect on the meat mix, resulting in easier extrusion and stuffing, and increased water binding results in improved yield and longer shelf life. It can also provide retort stability to canned pet foods; in those containing a gravy, it may provide viscosity.

Like guar gum, LBG can be used in barbecue sauces, salad dressings, and dry-mix soups and gravies. It can provide mouthfeel to marshmallows and meringue toppings and control water adsorption and dough and batter rheology in the production of bread, cake, and biscuits. It improves lubrication and water retention in tortillas and tacos and enhances the crispness of rehydrated freeze-dried products such as celery.

Products

Typical products containing guar and locust bean gums are given in Table 9.1. Guar gum is used more widely and extensively than is LBG because of its lower cost and easier dissolution, in addition to the difference in properties.

Appropriate label designations are *guar gum* and *locust bean gum* or *carob gum*.

Chapter 10

Xanthan

Microorganisms produce polysaccharides as structural elements and protective coatings to ward off invasive organisms and/or to prevent excessive moisture loss under drying conditions. A polysaccharide coat or capsule usually is produced when the food supply or nutrients become less available or as the water supply diminishes. Capsular material is composed of high-molecular-weight polysaccharides as a gel layer or, under drying conditions, a cohesive film that, on later contact with water, rehydrates to a thick gel coating. Capsules do not develop to significant extents when the organisms have available to them an ample supply of nutrients and are undergoing rapid multiplication in the logarithmic growth phase. Under proper conditions, *Xanthomonas campestris*, a bacterium commonly found on leaves of plants of the cabbage family, produces a polysaccharide, termed *xanthan*, that encapsulates the cell and diffuses into the surrounding medium.

Xanthan is widely used as a food gum. It is produced commercially by a pure-culture, submerged, aerobic fermentation. The characteristics of xanthan vary with variations in the strain of the organism, the sources of nitrogen and carbon, the degree of medium oxygenation, the temperature and pH of the fermentation, and the concentrations of various inorganic ions. When the fermentation is finished, the broth is pasteurized to kill the organisms, and the gum is recovered by precipitation with isopropanol. The polysaccharide is known commercially as *xanthan gum*.

Structure

The backbone chain of xanthan molecules has the same structure as that of cellulose molecules. In the xanthan molecule, every other β-D-glucopyranosyl unit in the cellulose backbone has attached, at the O-3 position, a β-D-mannopyranosyl-(1→4)-β-D-glucuronopyranosyl-(1→2)-6-O-acetyl-β-D-mannopyranosyl trisaccharide unit (Fig. 10.1).[1] About half of the terminal β-D-mannopyranosyl units have pyruvic acid attached as a 4,6-cyclic acetal (Fig. 10.2). Close alignment of the trisaccharide side chains with the main chain make the molecule a rather stiff rod (Fig. 10.3) with extraordinary stability

---

[1] Bacterial heteroglycans usually have regular, repeating-unit structures, distinguishing them from plant heteroglycans, which do not.

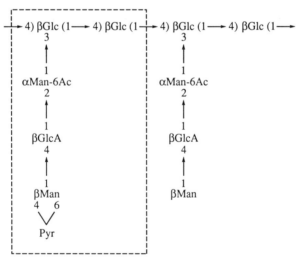

Fig. 10.1. Structure of the pentasaccharide repeating unit of xanthan, actual (top) and shorthand (bottom) designation. Inside the box is a pyruvylated pentasaccharide building block unit as in the upper structure at the top; outside the box is a nonpyruvylated unit.

to heat, acid, and alkali. The molecular weight is probably in the order of 2 million, although much larger values, presumably due to aggregation, have been reported. Xanthans of the highest pyruvic acid content have the highest viscosities and thermal stabilities.

Properties

Because of their stiffness, xanthan molecules are extended in solution. Therefore, solutions of xanthan have high viscosities and are highly pseudoplastic (Chapter 5). These characteristics are exhibited over a wide range of pH values and a temperature range

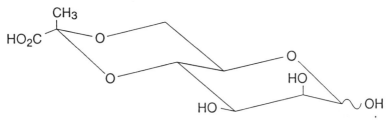

Fig. 10.2. The 4,6-*O*-pyruvyl-D-mannopyranosyl nonreducing end-unit of a repeating unit of xanthan.

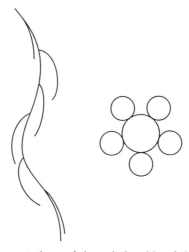

Fig. 10.3. Two representations of the relationship of the trisaccharide side chains to the backbone helix of xanthan molecules. The left-hand view is from the side. The right-hand representation is a view down the helix axis; the small circles represent the positions of the trisaccharide side chains and the large central circle the backbone helix.

from 0 to 100°C, that is, from freezing to boiling conditions. Xanthan solutions have a Newtonian plateau at lower shear stress values (Fig. 10.4), which means that the viscosity of xanthan solutions is not reduced by very low rates of shear. This makes it excellent for generating and stabilizing emulsions and suspensions, for example, suspensions of relishes and herbs in salad dressings.

Xanthan interacts with galactomannans, such as guar and locust bean gum, giving a synergistic increase in solution viscosity. The interaction with locust bean gum produces a heat-reversible gel (Fig 9.4). It is possible that the interaction also occurs with single helical xanthan (Fig. 10.5; compare to Fig. 9.2).

Uses

Xanthan is useful as a food gum because of its solubility in hot or cold water, high solution viscosity at low concentrations, little to no change in solution viscosity in the temperature range from 0 to

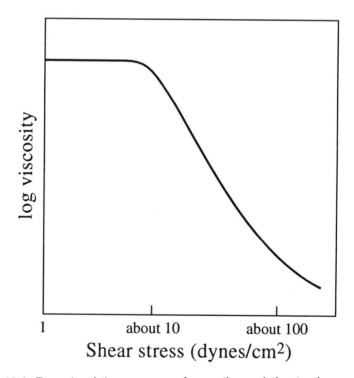

Fig. 10.4. Example of the response of a xanthan solution to shear rates, showing the Newtonian plateau and the pseudoplastic region.

100°C (which makes it unique among food gums; see Chapter 5), excellent thermal stability, solubility and stability in acidic systems, a high degree of solution pseudoplasticity (Chapter 5), a Newtonian plateau at low shear values, excellent compatibility with salt, and good solution freeze-thaw stability. The unusual and very useful properties of xanthan undoubtedly result from the structural rigidity of its molecules, which in turn results from its linear, cellulosic backbone that is stiffened and shielded by the trisaccharide side chains.

Xanthan is ideal for stabilizing aqueous dispersions, suspensions, and emulsions. The fact that the viscosity of its solutions changes indiscernibly with temperature, that is, its solutions do not thicken upon cooling, make it irreplaceable for thickening and stabilizing such things as pourable salad dressings and chocolate syrup, which need to pour as easily when they are taken from the refrigerator as they do at room temperature, and useful for making gravies, which should neither thicken appreciably as they cool nor thin too much on a steam table. In regular salad dressings, it is at the same time a thickener and a stabilizer for both the oil-in-water emulsion and the suspension of any particles, such as spices. It is an effective stabilizer of both suspensions and emulsions because its solutions have the Newtonian plateau. In no-oil (reduced-calorie) dressings, it thickens and suspends. In both oil-containing and no-oil salad dressings, xanthan is almost always used in combination with propylene glycol alginate (PGA) (Chapter 12). PGA produces solutions of much lower viscosity and much less pseudoplasticity. So

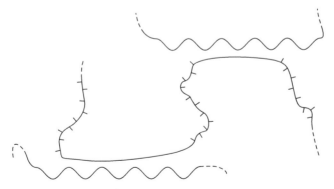

Fig. 10.5. Representation of the hypothesized interaction of a locust bean gum molecule with single helical portions of xanthan molecules to form a three-dimensional network and a gel.

when the two gums are used together, they produce a solution of less viscosity than would be realized if only xanthan were used and give the pourability associated with the pseudoplastic xanthan and the creaminess sensation (viscosity at the shear rates encountered in the mouth) associated with a nonpseudoplastic solution. They are generally used in a xanthan-PGA ratio of about 1:2, with 0.6% PGA and 0.3% xanthan allowing the oil content to be reduced partially or all the way to zero. Other products in which it finds common use are given in Table 10.1.

TABLE 10.1
Typical Products Containing Xanthan Gum

Bakery products	Mixes
Cake mixes	Cake mixes
Danish fillings	Cocktail mixes
Pie crust	Fruit drink mixes
Pie filling mixes	Gravy mixes
Poptarts	Pie filling mixes
Refrigerated doughs	Pudding mixes
Cereal bars	Salad dressings, etc.
Condiments	Low-calorie pourable salad
Pickle relish	dressings
Salsa	Pourable salad dressings
Dairy products	Reduced-calorie mayonnaise
Cheesecake	Sauces
Cheese spreads	Barbecue sauces
Cottage cheese	Cocktail sauces
Cream cheese	Mushroom sauce
Frozen cheese	Oriental sauce
Sour cream	Pizza sauces
Whipped toppings	Sloppy Joe Sauce
Egg substitutes	Taco sauces
Frozen foods	Tartar sauce
Frozen cheese	Spreads
Frozen fish florentine	Margarine spreads
Frozen guacamole	Sandwich spreads
Frozen lasagna	Syrups
Frozen pizza	Chocolate syrup
Frozen rice	Pancake syrup
Honey-roasted peanuts	Toppings
Meat products	Butterscotch topping
Breakfast slices	Fudge topping
Poultry breast slices	Marshmallow creme
	Whipped toppings

Blends of xanthan and locust bean gum and/or guar gum are excellent ice cream and frozen novelty stabilizers (see Chapters 9 and 11). Carrageenan is also added to the mix to prevent whey separation during freezing (Chapter 11). Other applications of these two- or three-gum blends are in the preparation of canned frostings, chip dips, cream cheese, cottage cheese and cottage cheese products, processed cheese products, and pizza sauce.

Ingredient labels for products containing xanthan may read *xanthan* or *xanthan gum*.

Chapter 11

Carrageenans

Carrageenans are mixtures of several related galactans having sulfate half-ester groups attached to the sugar units. They are obtained from red seaweeds, family Rhodophyceae. Principal sources are *Chondrus crispus,* obtained along the North American coast from Boston to Halifax and Prince Edward Island, and *Eucheuma* species, obtained mainly by farming in shallow waters around the Philippines, where 60% of the world production of this weed is obtained. Carrageenans constitute 30–80% of the cell walls of these marine algae, the amount depending on species, season, and growing conditions.

Carrageenans are obtained by extraction of the appropriate seaweed or seaweeds with hot, dilute alkaline solution, which produces the normal commercial product, sodium carrageenate. The product is recovered by precipitation with isopropanol; for nonfood uses it may be obtained by drum drying or by freezing. The product is polymolecular, that is, a mixture of molecules with similar but different structures, and is simply called *carrageenan* without regard to source or principal structure.

Also available and used is a product called processed Euchema seaweed (PES), Philippine natural grade (PNG) carrageenan, or alkali-modified seaweed flour. To prepare PES/PNG carrageenan, red seaweed, primarily *Eucheuma* species, is treated with $2M$ potassium hydroxide. Because the potassium salts of the types of carrageenans found in these seaweeds are insoluble (see Properties), the polysaccharide is not removed. Primarily, low-molecular-weight soluble components are removed from the plants during the treatment. The remaining dried seaweed is ground to a powder. PES/PNG carrageenan is therefore a composite material that contains not only the molecules of carrageenan that would be extracted out with dilute sodium hydroxide or sodium carbonate solution, but also other cell wall materials.

Two other food gums, agar and furcellaran (also called *Danish agar*), are also extracted from red seaweeds and are closely related structurally and in properties to the carrageenans.

Structures

The term *carrageenan* denotes a group or family of sulfated galactans extracted from red seaweeds, rather than a single polysaccharide. Carrageenan molecules are linear chains of D-galactopyranosyl units joined with alternating (1→3)-α-D- and (1→4)-β-D-glycosidic linkages, with most sugar units having one or two sulfate groups esterified to a hydroxyl group at carbon atoms C-2 or C-6. Sulfate contents range from 15 to 40% by weight. Units often contain a 3,6-anhydro ring. The principal structures are designated kappa (κ), iota (ι), and lambda (λ) (Fig. 11.1).

The repeating disaccharide units shown in Figure 11.1 represent the predominant building block of each type. Carrageenans, as extracted, are mixtures of these nonhomogeneous polysaccharides. Under alkaline extraction conditions, and especially if strong alkaline solutions are used, some λ-carrageenan is converted into κ- or ι-carrageenan, due to nucleophilic displacement of some sulfate groups at the C-6 position by hydroxyl oxygen atoms at C-3 positions to create 3,6-anhydro rings.

Properties

Carrageenan products, of which there may be more than 100 for different specific applications from a single supplier, contain different proportions of the three main behavioral types: λ, κ, and ι. The composition and properties of carrageenan preparations depend upon the species collected (or grown), growth conditions, and treatment during production. Carrageenan preparations are then blended and standardized to provide products for a variety of specific applications (see Uses). Alternatively, seaweeds are blended before extraction to provide products with specific properties. Properties that can be controlled by processing include dispersibility, rate of hydration, gel strength, protein interaction, and viscosity.

Carrageenan products dissolve in water to form highly viscous solutions with pseudoplastic behavior. The viscosity is quite stable over a wide range of pH values because the sulfate half-ester groups are always ionized, even under strongly acidic conditions, giving the molecules a negative charge. Coulombic repulsion maintains the molecular segments in locations maximally removed from each

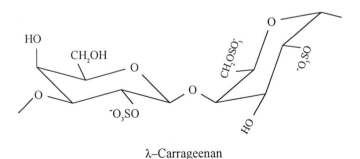

Fig. 11.1. Idealized repeating unit structures of kappa (κ), iota (ι), and lambda (λ) types of carrageenans.

other, which tends to produce a linear arrangement with the overall configuration of an expanded coil. Coulombic repulsion also prevents interchain segmental collisions. Nevertheless segments of molecules of κ- and ι-type carrageenans can, and do, exist as double helices of parallel chains formed from right-handed, threefold helical stretches. Addition of potassium or calcium ions to a hot solution containing such double-helical segments, followed by cooling, produces thermoreversible gels. Gelation, which can occur in water at concentrations as low as 0.5%, takes place instantaneously at the gelling temperature in water, but gel setting may require several hours.

κ-Types gel most strongly with potassium ions. When 4% κ-type carrageenan solutions are cooled below 45–60°C (115–140°F) in the presence of potassium ions, a rigid, brittle gel is formed. Calcium ions are less effective in producing gelation, but potassium and calcium ions together produce a high gel strength. κ-Type gels are the strongest and stiffest of the carrageenan gels but, like others, tend to synerese as junction zones extend within the structure, forcing water out of the gel (see Chapters 4 and 5). The presence of other gums retards syneresis. Only the sodium salts of κ-type carrageenans are soluble in cold water.

ι-Types are a little more soluble than are the κ-type carrageenans, but as with the κ-types, only the sodium salt form is soluble in unheated water. ι-Types gel best with calcium ions. The resulting gels are soft and flexible (elastic or resilient), have good freeze-thaw stability, and do not synerese, presumably because ι-type carrageenans are more hydrophilic and produce fewer junction zones than do κ-type carrageenans. ι-Type carrageenans exhibit synergistic interactions with starches.

Gelation occurs because, as solutions of κ- or ι-type carrageenans are cooled, the linear molecules form double or triple helices that are not continuous due to the presence of structural irregularities. The linear helical portions then associate to form a rather firm, three-dimensional, stable gel in the presence of the appropriate cation (Fig. 11.2). All salts of λ-type carrageenans are soluble and nongelling.

Under conditions in which double-helical segments occur, carrageenan molecules, particularly those of the κ type, form junction zones with the naked segments of locust bean gum (LBG) to produce rigid, brittle, syneresing gels (Fig. 9.2). Only one-third as much of the gum mixture is needed to form a gel equivalent to that formed using a pure κ-type carrageenan.

Carrageenans are quite susceptible to acid-catalyzed hydrolysis, undergoing rapid and extensive loss of viscosity when solutions below pH 5 are heated. This is primarily due to cleavage at the (1→3) glycosidic linkages, except when an O-2 sulfate group is present.

Uses

Commercial carrageenans can be viscosity builders, gelling agents, or stabilizers. Carrageenan products are most often used because of their ability to form gels with milk and water (Table 11.1). Blending provides a wide range of products that are standardized with various amounts of sucrose, glucose (dextrose), buffer salts, or gelling aids, such as potassium chloride. The

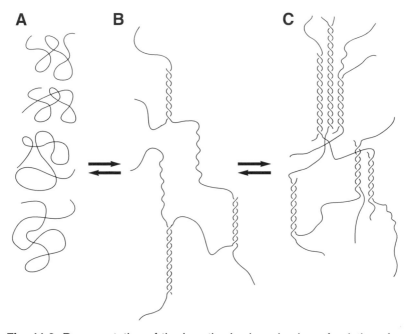

Fig. 11.2. Representation of the hypothesized mechanism of gelation of κ- and ι-type carrageenans. The polymer molecules in a hot solution are in a coiled state (A). As the solution is cooled, they intertwine in double helical structures (B). Upon further cooling, the double helices are believed to nest together with the aid of potassium or calcium ions (C).

available commercial products form a variety of gels: gels that are clear or turbid (with water), rigid or elastic, tough or tender, heat-stable or thermally reversible, and that do or do not undergo syneresis. Carrageenan gels do not require refrigeration because they do not melt at room temperature. They are also relatively freeze-thaw stable.

A useful property of carrageenans is their reactivity with proteins, particularly those of milk. κ-Type carrageenans complex with κ-casein micelles of milk, forming a weak, thixotropic, pourable gel. The thickening effect of κ-carrageenans in milk is 5–10 times greater than it is in water. This property is used in the preparation of chocolate milk, where the thixotropic gel structure prevents settling of the cocoa particles, thereby maintaining a uniform distribution. Such stabilization requires a gum concentration of only about 0.025%. This property is also applied in the preparation of ice cream, evaporated milk, infant formulas, freeze-thaw-stable whipped cream, and emulsions in which milk fat is replaced with a vegetable oil. Carrageenans are often used as a secondary stabilizer in ice cream to prevent whey separation during freezing. The κ- and ι-type carrageenans can gel milk at 0.1–0.2% concentration. In evaporated milk, carrageenans stabilize the α-casein and prevent creaming at concentrations as low as 0.005%.

ι-Type carrageenans are more soluble than are the κ-types, but

TABLE 11.1
Typical Products Containing a Carrageenan

Brownie mixes	Frozen novelties
Cheesecakes	Frozen pizza
Chocolate milk	Frozen yogurts
Chocolate milk mixes	Horseradish sauce
Chowders	Hot cocoa mixes
Complete pancake mixes	Ice creams
Cottage cheese	Instant breakfast drinks
Cream cheese	Low-calorie salad dressings
Dairy-based diet products	Low-fat yogurts
Dessert toppings	Non-dairy coffee creamers
Dips	Pressurized whipped creams
Egg substitutes	Pudding and pie filling mixes
Evaporated milk	Sour cream
Frozen cheese	Whipped toppings
Frozen cheese lasagna	

with both, only the sodium salt forms are soluble in cold water. While water gels of κ-type carrageenans are firm, rigid, and characterized by syneresis, ι-type carrageenans form elastic, syneresis-free, thermally reversible gels that are stable to repeated freeze-thaw cycling. They are used in ready-to-eat milk gels. Blending is common, and blends of ι- and κ-type carrageenans are used in water-dessert gels that do not require refrigeration (as opposed to gelatin gels), whipped toppings, instant whipped desserts, and eggless custards and flans. Gelled water desserts containing fruit concentrates or flavors, usually at a pH of about 3.8, contain 0.5–1.0% carrageenan. Acid is generally added just before rapid cooling. Carrageenan is often used in combination with LBG to yield a fast-setting, temperature-stable gel.

The synergistic effect between κ-carrageenan and LBG produces gels with greater elasticity, greater gel strength, and less syneresis than gels of potassium κ-carrageenate alone (see Chapter 9). As compared to κ-type carrageenan alone, the combination of κ-type carrageenan and LBG provides greater stabilization and air bubble retention (called *overrun*) in ice cream, but also a little too much chewiness, so guar gum is added to soften the gel structure (see Chapter 9). The LBG-κ-carrageenan combination is also used in canned pet foods, in yogurt to provide body and fruit suspension, and in low-fat yogurt for stabilization. Tart filling and tart glazing can be gelled with a mixture of a κ-type carrageenan (1–2%) and LBG (0.7–1%). High gelation and melting temperature is obtained on addition of a potassium salt. Freeze-thaw stability is improved by addition of carboxymethylcellulose.

In summary of what has been discussed in Chapters 7 and 9–11, most ice creams and related milk shake and frozen dessert products contain at least four gums. Carboxymethylcellulose (Chapter 7) is used as the primary stabilizer. κ-Type carrageenan (Chapter 11) is added as a secondary stabilizer to prevent whey separation. LBG (Chapter 9), through its interaction with the carrageenan, converts the mix to a weak gel structure. Guar gum (Chapter 9) is added to soften and make less chewy the gel structure formed with LBG and κ-carrageenan, providing a smooth texture and creamy body. Stabilization can also be achieved with blends of xanthan (Chapter 10), LBG, and guar gum, which give viscosity control in the mix and tolerance to heat shock. Reduced butterfat products generally contain microcrystalline cellulose (Chapter 7) and/or a protein- or starch-based fat mimetic (Chapter 15). Mixes containing three to five of these polysaccharides plus other ingredients, such as emulsifiers (mono- and diglycerides), are prepared and sold as a single component to be added to milk of various fat contents. Flavors may also be included or may only need to be added.

Cold hams and poultry rolls take up 20–80% more brine when they contain 1–2% of a κ-type carrageenan, and slicing is improved. Addition of 0.2–1.0% of carrageenan alone or in combination with LBG improves the gelled broths of canned meat and fish products. Carrageenan coatings can serve as a mechanical protection and a carrier for seasonings and flavors. Meat emulsions that contain κ- or ι-type carrageenans have improved adhesion and water content. Meat analogs from casein and vegetable proteins in fiber form may contain 1% carrageenan. A growing use of carrageenan is to hold water and thereby maintain the water content and softness of meat products, such as wieners and sausages, during the cooking operation. Addition of a κ- or ι-type carrageenan in the Na^+ form or PES/PNG carrageenan to low-fat ground beef maintains texture and hamburger quality. Normally, fat serves the purpose of maintaining softness, but because of carrageenan's power to bind to protein, it can be used to replace, in part, this natural animal fat functionality in lean products.

λ-Type carrageenans are nongelling gums. They are used as emulsion stabilizers in products such as whipped cream, instant breakfast drinks, imitation coffee creams, and milk shakes. A blend of κ- and λ-carrageenan is used in the latter.

Labels of products containing either κ-, λ-, or ι-type carrageenans may read *carrageenan, chondrus extract,* or *Irish moss extract.*

Agar Uses

The principal use of agar is in bakery icings and frostings because it is compatible with large amounts of sugar and its products neither melt at high storage temperatures, such as might be found in a delivery truck, nor stick to the packaging material. It can also be found in no-oil salad dressings, light sour cream, and yogurt. Agar may be designated *agar* or *agar-agar.*

Chapter 12

Alginates

Commercial algin is a salt, usually the sodium salt, of alginic acid, a linear poly(glycuronic acid), which constitutes 18–40% of the dry weight of brown seaweeds, family Phaeophyceae. The major source of algin/alginates is the giant kelp, *Macrocystis pyrifera*, growing in coastal water from just north of San Diego south into Northern Mexico. This seaweed, attached to the rocky seafloor, grows to a length of more that 120 ft (37 m) and has carbon-dioxide-filled bladders of about an inch in diameter along the strand, which float it in an upright position so that it can receive sunlight. Under optimal conditions, fronds[1] will grow about 2 ft (about 60 cm) per day. Harvesting is done by a collector ship that moves through the floating kelp cutting the plants several feet (about 1 m) below the surface and continuously lifting the cut kelp by means of a conveyor into the ship. Another alginate source is seaweeds of the genus *Ascophyllum* that grow to a length of 3–5 ft (1–1.6 m) on rocky shores and are harvested by hand.

Harvested seaweed is washed. Treatment with a sodium carbonate solution converts the polysaccharide into its soluble sodium salt, which is removed by extraction. Alginate is then precipitated from the extract as its insoluble calcium salt, and the precipitate is washed free of solubles. Addition of hydrochloric acid converts the preparation into insoluble alginic acid. Finally, the alginic acid is neutralized and converted into a soluble monovalent ion salt, most often into sodium alginate.

Structure

Alginic acid is composed of β-D-mannopyranosyluronic acid and α-L-gulopyranosyluronic acid units. These two monomeric units occur in

---
[1] Fronds are leaflike branches.

homogeneous regions composed of only one of the two units and in regions containing both units. Segments containing only D-mannuronopyranosyl units are referred to as M blocks and those containing only L-guluronopyranosyl units as G blocks. MG blocks contain the two units in a mixed arrangement. D-Mannuronopyranosyl units are in the $^4C_1$ conformation, while L-guluronopyranosyl units are in the $^1C_4$ conformation, which gives the different blocks quite different chain configurations (Figs. 12.1–12.3). Different percentages of the different block segments cause alginates from different seaweeds to have different properties. Alginates with high G-block contents produce gels of high strength (see Properties). Average degrees of polymerization from 80 to 1,000 have been reported.

In a bacterial alginate produced by *Azotobacter vinelandii*, about 82% of its sugar units alternate along the chain, resulting in poor gelling properties.

Propylene glycol alginates (PGAs) are made by reacting moist alginic acid with propylene oxide, causing 50–85% of the carboxyl

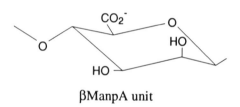

βManpA unit

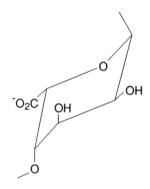

αLGulpA unit

Fig. 12.1. Units of β-D-mannopyranosyluronic acid (βMan*p*A) in the $^4C_1$ conformation and α-L-gulopyranosyluronic acid (αLGul*p*A) in the $^1C_4$ conformation.

groups to become esterified. In the reaction below, R-COO⁻ represents a unit of D-mannopyranosyluronic acid or L-gulopyranosyluronic acid.

$$R-\underset{\underset{O}{\|}}{C}-O^- + CH_2-CH-CH_3 \text{ (epoxide)}$$

$$\downarrow$$

$$R-\underset{\underset{O}{\|}}{C}-O-CH_2-CHOH-CH_3$$

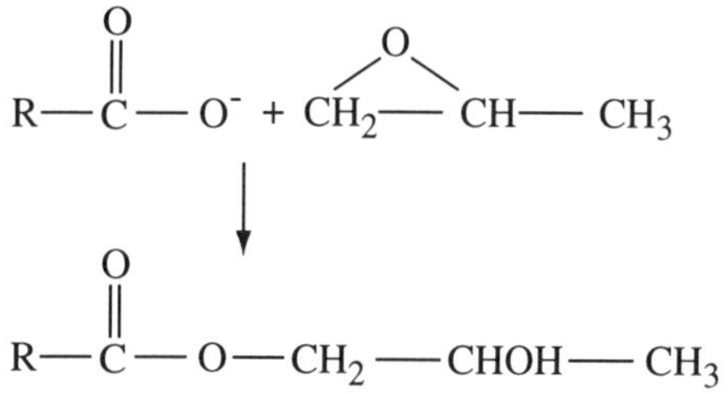

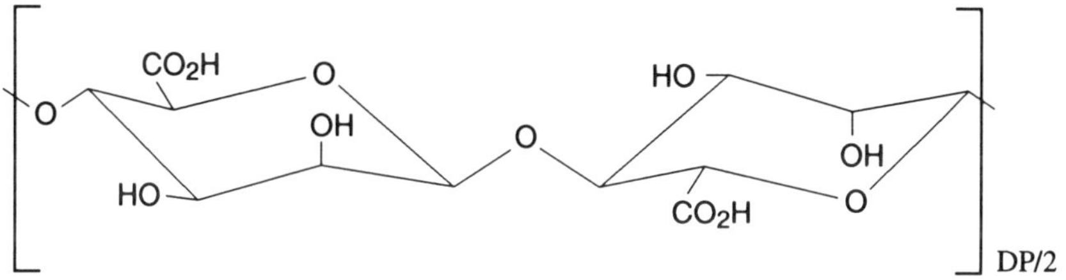

Fig. 12.2. Repeating unit segment of an M-block region. Note the similarity of this structure to that of cellulose (Chapter 4) and that of the mannan backbone of the galactomannans (Fig. 9.1) and the flat, ribbonlike structure formed by the equatorial→equatorial glycosidic bonds.

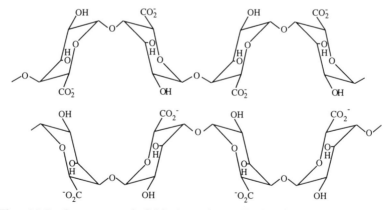

Fig. 12.3. Segments of G-block regions. Each of the two chains is composed of two repeating units. Note the pleated conformations of the chains formed by the axial→axial glycosidic bonds and the "holes" that complex calcium ions. Compare with Figure 13.3.

Properties

Alginate solutions are slightly pseudoplastic. Very small amounts of calcium ions will increase viscosity. Slightly more will make alginate solutions thixotropic. And more will convert them into permanent gels.

Sodium alginate solutions gel upon addition of calcium ions, which fill the cavities formed between parallel G-block chain regions (Fig. 12.4.). These cavities contain two carboxylate and two hydroxyl groups, one each from each chain. The result is a junction zone that has been called an "egg box" arrangement, with the calcium ions being likened to eggs in the pockets of an egg carton (Fig. 12.5.). Strength of the gel depends on the content of G blocks in the alginate used and the concentration of calcium ions. Because of the effect of cross-linking of alginate

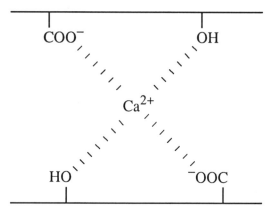

Fig. 12.4. Coordination of a calcium ion with hydroxyl and carboxyl groups on adjacent G-block chains.

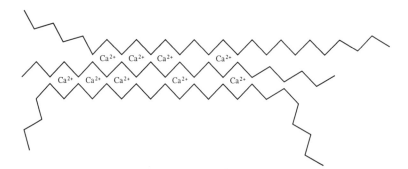

Fig. 12.5. Representation of the proposed formation of a junction between two alginate molecules promoted by calcium ions.

molecules by polyvalent ions, such as calcium ions, the hardness of water used to dissolve algin has an important effect on solution viscosity.

Since alginate molecules are salts of weak carboxylic acids, with pK_a values of 3.38 (ManA) and 3.65 (LGulA), the extent of carboxyl group ionization decreases as pH is decreased and is strongly repressed at pH values of 3 or less. This loss of ionization at low pH values causes the alginate chains to lose their negative charge, giving the molecules more of the characteristics of long neutral molecules, and allows junction zones to develop. As molecular associations increase, the viscosity of the solution increases. Finally, when the pH is low enough, alginic acid precipitates from solution because of extensive intermolecular associations.

Alginic acid, although insoluble, swells in water and is used as a tablet disintegrant. It is sensitive to heat, a property that is characteristic of all polysaccharides with free carboxylic acid groups, for example, carboxymethylcellulose (Chapter 7).

Solutions of PGA are thixotropic and much less sensitive to low pH values and polyvalent cations, including calcium ions and proteins, than are solutions of nonesterified alginates. Because esterified carboxyl groups cannot ionize, PGAs have much less negative charge than do salts such as sodium alginate. However, the propylene glycol group introduces a "bump" on the chain that prevents chain association, giving stability to PGA solutions. Because PGAs can tolerate calcium ions, they can be used in dairy products. The hydrophobic propylene glycol groups also give the molecule surface activity, that is, foaming, emulsifying, and emulsion-stabilizing properties.

Uses

Algin is widely used to provide viscosity at low concentrations. Higher viscosity (under low shear conditions) can be achieved by adding a small amount of calcium ions to the gum solution. This characteristic is used to thicken fruit juices and to provide suspension to juice insolubles. Thickening action, and particularly thixotropic behavior brought about by low calcium ion levels, tends to hold the pulp in suspension. When PGA is used, the calcium ion cross-linking of chains that occurs through the remaining carboxylate groups produces thickening without producing a weak thixotropic gel. The mild interfacial activity of PGA, which is due to the hydrophobic character of the ester groups, stabilizes flavor oil emulsions. Typical products made with algins are shown in Table 12.1.

Calcium alginate gels are obtained by diffusion setting, internal setting, and setting by cooling. Diffusion setting can be used to prepare structured foods. Two examples are structured pimento strips and structured onion rings. In the production of pimento strips for stuffing green olives, pimento puree is mixed with water containing a small amount of guar gum, which thickens the diluted puree and reduces shrinkage of the final strip. Then, sodium alginate is added. The mixture is applied to a conveyor belt as a thick film or sheet. Spraying the sheet with calcium chloride solution sets the surface. The moving belt is then passed through a calcium chloride setting bath, where more calcium ions diffuse into the sheet, gelling it even more. The final sheet is tough,

Table 12.1
Typical Products Containing Algins

Sodium alginate[a]

Alfredo sauce
Breakfast bars
Caramels
Cereal bars
Dry soup mixes
Frozen gravy
Frozen guacamole
Fruit purees
Nonrefrigerated dessert gels
Pimento pieces for Spanish olives
Sauce mixes
Structured fruit pieces
Structured onion rings

Propylene glycol alginate (PGA)

Buttered pancake syrups
Cheesecakes
Chocolate milk mixes
Cocktail mixes
Fruit punches
Icings
Low-pH syrups
Regular and reduced-calorie, pourable salad dressings
Sandwich spreads
Tartar sauce

Sodium alginate + acid

Fruit-containing, filled breakfast cereal products
Jelly-type bakery fillings
Nonrefrigerated dessert gels
Tomato aspic

[a] Many of these applications require natural or added calcium ions.

flexible, and resilient. Equilibration for several days in a solution of sodium chloride and calcium chloride allows the alginate gel to undergo syneresis and shrink to an acceptable level of consistency. The sheet is then cut into strips, which are bent double and stuffed into green olives by machine.

Structured, frozen onion rings are made in much the same way from minced onions or rehydrated onion flakes, sodium alginate, salt, and flavor. The resulting paste is formed into rings and gelled by immersion in a setting bath of 3–5% calcium chloride or by spraying with a calcium chloride solution. After setting, the rings are battered and fried.

Internal setting for fruit mixes, fruit purees, and structured fruit (apple, peach, pear, and apricot) pieces involves slow release of calcium ions within the mixture. In the preparation of fruit pieces, a two-mix system is employed. One mix contains sodium alginate and an insoluble calcium ion source such as dicalcium phosphate or calcium sulfate dihydrate. No reaction between the alginate and calcium ions occurs because the calcium salt is insoluble. The second mixture contains fruit, sequestrant, and acid. The two mixtures are then mixed and allowed to stand under shear-free conditions in their final container or shaped form on a conveyer belt. Setting occurs due to the controlled release of calcium ions by the acid and the normal partial crosslinking reaction of alginate chains. Dicalcium phosphate is generally not used on continuous conveyers because the faster release of calcium ions causes too rapid gel development.

Setting by cooling involves dissolving a calcium salt, a slightly soluble acid, and a sequestrant in hot water and allowing the mixture to set on cooling. In this process, calcium ions are released at a moderate rate when the mixture is heated, bringing about a low degree of crosslinking. Sufficient thermal energy is provided by the hot water that the alginate chains form only weak, temporary junction zones, and gelation does not occur until cooling slows thermal action. Gels produced in this way exhibit little or no syneresis, allowing the process to be used to make structured fruit pieces. The gels are also heat stable, so those containing fruit can be used for pie fillings that remain stable through pasteurization and cooking.

Unlike gelatin gels, alginate gels are not thermoreversible and can be used as dessert gels in warm climates. Alginic acid, that is, an alginate solution with lowered pH, with or without addition of some calcium ions, is employed in the preparation of soft, thixotropic, nonmelting gels.

Strong complexes and precipitates are often formed when the pH of a mixture containing proteins and algin is lowered from 7 to 5. Interaction between protein and alginate molecules occurs at pH values below the protein's isoionic pH value (i.e., its isoelectric point [pI]), where the protein molecules have a net positive charge. Alginate molecules remain negatively charged at pH 5. Alginate-whey and alginate-minced fish complexes are formed in this way.

Ingredient labels for products containing an algin may read *algin, alginic acid, sodium alginate, potassium alginate*, or *ammonium alginate* depending on the product used.

PGA is used when stability to acid, nonreactivity with calcium ions (for example, in milk products), or its surface-active property is desired. Accordingly, it finds use as a thickener in salad dressings, including low-calorie salad dressings. In both regular and low-calorie pourable dressings, it is often used in conjunction with xanthan (see Chapter 10), as PGA provides smoother flow than does xanthan alone. *Propylene glycol alginate* and *algin derivative* are acceptable label designations.

Chapter 13

Pectins

Commercial pectins are galacturonoglycans, i.e., poly(α-D-galactopyranosyluronic acids), with various contents of methyl ester groups. Native pectins are more complex molecules found in cell walls and intercellular layers of all land plants. Commercial pectin is obtained by acid extraction of citrus peel (20–30% pectin) and apple pomace (10–15% pectin), both by-products of juice manufacturing. That from lemon and lime peel is the highest quality pectin. For pectin production, citrus peel is extracted with water of pH 1.5–3.0 at 60–100°C (140–212°F). The extract is filtered, and pectin is precipitated by addition of isopropanol. In other countries, a small amount of pectin is produced from sunflower heads.

Structures

Compositions and properties of pectins vary with source, the processes used for handling and drying of the peel, the type of extraction, and subsequent treatments. During extraction with mild acid, some hydrolytic depolymerization and hydrolysis of methyl ester groups occur. Therefore, *pectin* denotes a family of compounds, which is part of a larger family of "pectic substances." Pectic acids, pectinic acids, and pectins are all pectic substances. *Pectic acids* refers to galacturonoglycans without, or with only a negligible content of, methyl ester groups (Fig. 13.1) and various degrees of neutralization. Salts of pectic acids are called *pectates*. *Pectinic acids* are galacturonoglycans with various, but greater than negligible, contents of methyl ester groups and may have varying degrees of neutralization. Salts of pectinic acids are *pectinates*. Pectins are mainly pectinic acids. The term *pectin* is usually used in a generic sense to designate water-soluble galacturonoglycan preparations of varying methyl ester contents and degrees of neutralization that are capable of forming gels. Some of the carboxyl groups of pectin are in the methyl ester form; some in the free acid form; and some in a sodium, potassium, or ammonium salt form, most often the sodium salt form. Ratios of the various forms of the carboxyl groups of the D-galacturonic acid units are determined by the source and the time, temperature, and concentration of reagents used in production.

Pectins are subdivided according to their degree of esterification (DE), which is the percentage of carboxyl groups esterified with methanol. Pectins with DE >50%, that is, with more than one-half of the carboxyl groups in the methyl ester form (-COOCH$_3$), are high-methoxyl (HM) pectins, more properly high methyl-esterified pectins; those with DE <50% are low-methoxyl (LM) pectins. In both cases, the remainder of the carboxyl groups are present as a mixture of free acid (-COOH) and salt (-COO$^-$Na$^+$) forms. Degree of amidation (DA) indicates the percentage of carboxyl groups in the amide form. The DE and DA values strongly influence the solubility, gel-forming ability, conditions required for gelation, gelling temperature, and gel properties of the preparation.

Treatment of a pectin preparation with ammonia dissolved in methanol converts some of the methyl ester groups into carboxamide groups. In the process, LM pectin (by definition) is formed. Amidated LM pectins may have 15–25% of the carboxyl groups converted into carboxamide groups. Table 13.1 shows the carboxyl groups in the three types of pectin.

Acetyl ester groups are sometimes attached to O-2 or O-3 positions and are especially numerous in sugar beet and sunflower seed pectins. They inhibit gelation.

The key feature of all pectin molecules is a linear chain of (1→4)-linked α-D-galactopyranosyluronic acid units. Pectin is, therefore, an

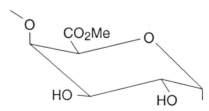

Fig. 13.1. The most prevalent monomeric unit of a high-methoxyl pectin.

α-D-galacturonan, i.e., a poly(α-D-galactopyranosyluronic acid) or an α-D-galacturonoglycan (Fig. 13.2). Neutral sugars, primarily L-rhamnose, are also present. In at least citrus, apple, and sunflower pectins, α-L-rhamnopyranosyl units seem to be inserted into the backbone chain at rather regular intervals. At least some pectins contain convalently attached, highly branched arabinogalactan chains and/or short side chains composed of D-xylosyl units on the rhamnogalacturonoglycan backbone. The inserted L-rhamnopyranosyl units may provide the irregularities in structure that limit the size of the junction zones and effect gelation. The presence of side chains may also limit the extent of chain association. Junction zones are formed between regular, unbranched pectin chains when the negative charges on the carboxylate groups are removed by addition of acid, when hydration of the molecules is reduced by addition of other solubles to a solution of HM pectin, and/or when pectinate polymer chains are bridged by multivalent cations such as calcium ions.

Sodium and calcium pectates, pectic acid, and pectinic acid occur in the solid state as right-handed helices. In solid pectinic acid, the polymer molecules pack in a parallel arrangement; pectates pack as corrugated sheets of antiparallel chains.

TABLE 13.1
Types of Pectins

High-Methoxyl (HM)	Low-Methoxyl (LM)	Amidated LM
-COOCH$_3$ (>50%)	-COOCH$_3$ (<50%)	-COOCH$_3$ (<50%)
-COOH	-COOH	-COOH
-COO$^-$ Na$^+$	-COO$^-$ Na$^+$	-COO$^-$ Na$^+$
		-CONH$_2$ (15–25%)

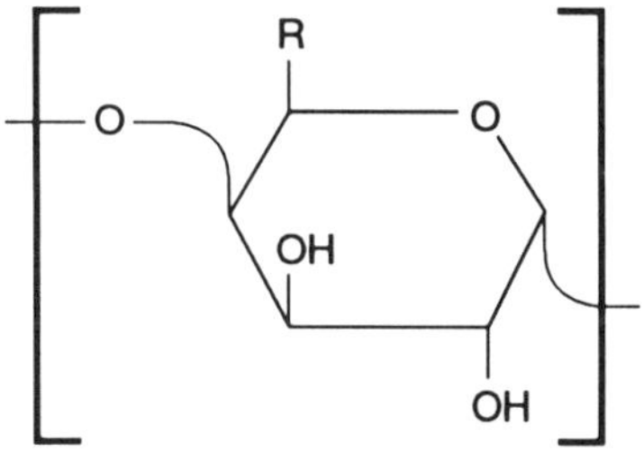

$$R = {-CO_2CH_3},\ {-CO_2}^-,\ or\ {-CO_2H}$$

Fig. 13.2. Representative segment of a nonamidated α-D-galacturonan (i.e., pectin) molecule.

Junction zones in HM pectin plus sucrose gels are believed to be formed by a columnar stacking of methyl ester groups to form cylindrical hydrophobic areas parallel to the helix axes. The same general "egg box" model used to describe the formation of calcium alginate gels (Chapter 12) is used to explain gelation of solutions of LM and amidated LM pectins upon addition of calcium ions. This model is used because of the close similarity in molecular structures of the (1→4)-linked poly(α-D-galactopyranosyluronic acid) segments of pectic acids (Fig. 13.3) and the (1→4)-linked poly(α-L-gulopyranosyluronic acid) segments of alginic acids, segments that are mirror images of one another except for the configuration at C-3.

Properties and Uses

Most pectin is used for making jams and jellies. However, amounts used in confections, beverages, and acidified drinks are increasing (Table 13.2). It is well suited for acidic foods because of its excellent stability at low pH values.

Solutions of HM pectin molecules gel when sufficient sugar and acid is present to remove some of the water of hydration. As the pH of a pectin solution is lowered, the highly hydrated and charged carboxylate groups are converted into uncharged, less-hydrated carboxylic acid groups. Upon losing some of their charge and hydration, segments of the polymer molecules associate, forming junctions and a network of polymer chains that entraps portions of the aqueous solution. Dehydration and junction zone formation is assisted by the

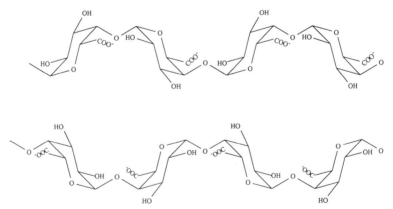

Fig. 13.3. Pectate-type segments of nonamidated low-methoxyl pectin showing the cavities that bind calcium ions. Compare with Figure 12.3.

presence of a high concentration (~65%, at least 55%) of a solute (sugar) that competes for water.

LM pectin solutions gel only in the presence of divalent cations, which provide cross-bridges. Increasing the concentration of divalent cations (only calcium ion is used in food applications) increases the gelling temperature and gel strength (Fig. 13.4). Because its solutions do not require sugar for gelation, LM pectin is used to make dietetic jams, jellies, and marmalades.

A solution of pectin of the required concentration, which also contains the required concentration of sugar and is at a proper pH value, prepared at a temperature above the gelling temperature and then cooled to decrease the thermal energy of the polymer molecules, will gel because the polymer molecules associate and form junction zones when they collide under these conditions. The temperature at which gelation occurs is the gelling temperature.

TABLE 13.2
Typical Products Containing Pectin

Cheese spreads	Icings and frostings
Cupcake mixes	Jams
Fillings for chocolates	Jellies
Fruit juices	Low- and no-sugar fruit spreads
Fruit-milk drinks	Low-calorie salad dressings
Fruit preserves of fruit yogurts	Marmalades
Fruit roll ups	Preserves

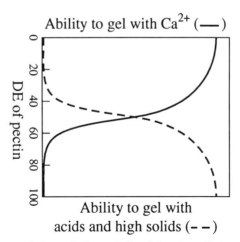

Fig. 13.4. Representation of the relationship of methyl ester content to conditions required for gelation. DE = degree of esterification.

Pectin gels are evaluated on the basis of strength within certain elastic limits or the actual breaking strength as measured by detecting the force needed on a plunger to break the surface of a gel. Sag value, also a measure of quality, is the decrease in height of a gel during a designated period after removal from a container of defined size.

Setting of pectin gels depends upon pectin composition, pH, temperature, and the concentration of water-binding components (primarily sugar) present. Sugar competes for water in the solution and reduces hydrogen bonding of water to the pectin. This lowered solvation allows more intermolecular bonding, or junction zone formation, and promotes gel formation.

Various gelling grades of HM pectin depend upon methyl ester content. Rapid-set pectins have a 72–75% methoxyl content (DE 72–75) and set within 20–70 sec at pH 3.0–3.1. Setting occurs with 0.3% pectin and about 65% soluble solids (sugar). Slow-set pectins have DE values of 62–68 and set in 180–250 sec. Medium-set pectins have a methoxyl content of 68–71%. All require an acidic medium for gelation, slow-set pectins requiring the lowest pH.

HM pectin gels are not usually remeltable, and no gel forms at pH values above 3.5 or sugar concentrations below 55%. The setting temperature can be varied from about 35°C (95°F) to about 90°C (195°F). Gels produced are rigid and hold a cut surface. Syneresis occurs. Calcium ions are not a factor in the setting of HM pectins. These pectins are good for making bakery jellies because they are pumpable and, when set, resist melting.

LM pectin gels can be remelted and reformed repeatedly. The remelt temperature can be as high as 150°C (300°F). These gels are spreadable, thixotropic, and generally shear reversible. LM pectins will set at pH values of up to about 6.5 (but the usual pH range is 1–5) and require no sugar for gelation since gel formation is dependent on the presence of Ca^{2+}. Calcium ions cause gel formation by cross-linking in the same way that calcium ions produce gels with alginate solutions (Figs. 12.4 and 12.5). They are always required for gelation of LM pectin solutions. Gelation will occur in the presence of 0–80% soluble solids, and the amount of sugar present affects the concentration of Ca^{2+} required. Increased calcium ion concentration increases gel strength and allows high gelling temperatures. (The setting temperature is generally in the range 40–100°C [105–212°F].) However, an increase in calcium ion concentration above the concentration required to saturate the polymer, that is, to crosslink all carboxylate groups, lowers the gel strength and setting temperature and may produce an undesirable "apple sauce" structure. The pH of a gel does not influence its texture.

LM pectins make good barbecue sauce because of their ability to provide cling.

Amidation is the introduction of carboxamide groups. It makes the LM pectin much more sensitive to calcium ions (Fig. 13.5), so less is required and the amount normally present in fruit suffices for strong gel production. In fact, amidated LM pectins (LMA) are so sensitive to Ca^{2+} that the normal calcium ion content of water can be a problem, because neither pectins nor any other "gelling" gum will dissolve in an aqueous system at a temperature at which the system would otherwise be gelled by the gum. At pH values less than 3.5, LMA pectin gels are similar to those made with HM pectin but are more rubbery. Above pH 3.4–3.5, LMA pectin gels are spreadable, thixotropic, and shear reversible. Their setting temperature is generally in the range 30–70°C (85–160°F). These gels are thermoreversible, with the remelt temperature generally being below 75°C (165°F). They can be used to make low-calorie jams and jellies because sugar is not required for gelation as it is with HM pectins.

Pectins are used primarily because of their unique ability to form

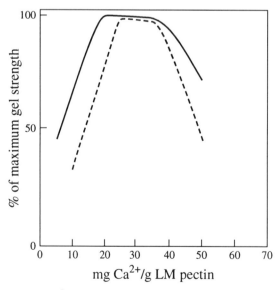

Fig. 13.5. Effect of Ca^{2+} ion concentration on the maximum gel strength of amidated (—) and nonamidated (---) low-methoxyl (LM) pectin.

spreadable gels in the presence of sugar and acid or in the presence of calcium ions. Rapid-set pectins are used in canned jams where quick solidification is needed to prevent flotation of fruit and to provide uniformity of product. Jellies require slow-setting pectin to provide time for bubbles to rise and escape.

Milk-fruit drinks can be stabilized by LM pectin during pasteurization because the negatively charged pectin combines with the milk protein, preventing agglomeration and settling. A similar behavior is seen with certain alginates (Chapter 12) and carboxymethylcelluloses (Chapter 7).

Small gel particles made with LM pectin produce some of the mouthfeel of fat, that is, they act as a fat mimetic (Chapter 15). Likewise, a pectin especially prepared by means of an acid extraction and processed into a fine powder swells in water to form small gel particles that have similar fat-replacing properties. Calcium-reactive pectin technology has also been used to reduce oil uptake by as much as 50% in french-fried potatoes and battered and breaded food products.

Both HM and LM pectin may be labeled *pectin*.

Chapter 14

Exudate Gums

When the bark of some trees and shrubs is injured by insects or by manual tapping, the plants exude a sticky substance that rather quickly hardens to seal the wound and give protection from infection and desiccation. Such exudates, called *tears*, are commonly found on plants that grow in semiarid conditions. These exudate gums were articles of commerce long before recorded history because they could be picked by hand and dissolved in water to produce a thick and generally adhesive paste. This paste was serviceable as a food ingredient, for medicinal applications, for sizing fabrics and early papers, and for printing. Gum arabic (gum acacia), gum karaya, and gum ghatti are tree exudates, and gum tragacanth is a shrub exudate. Although the exudate gums still have commercial markets, their overall use has, and continues to, diminish because of their uncertain availability and their increasing costs. Only gum arabic still has a good market in its traditional food applications.

All exudate gums require cleaning and pasteurization, since when freshly exuded they are sticky and trap dust, sand particles, insects, and/or bacteria. Pieces of bark that may remain attached to the tear when it is removed may provide a slight amount of color.

Gum Arabic

Gum arabic or gum acacia, the principal exudate gum used in food, has had a long and fascinating history. It was a commercial article at least 4,000 years ago, being shipped on Egyptian fleets and described in hieroglyphs as "kami" for use as an adhesive for mineral paints. It is an exudate of acacia trees, of which there are about 500 species distributed over tropical and subtropical areas of

Africa, India, Australia, Central America, and southwestern North America. The most important growing areas for the species that give the highest quality gum are the Sudan, where *Acacia senegal* is produced, and Nigeria, where several varieties of these Leguminosae are found.

The most important area for yielding gum arabic is central Sudan, where gum arabic is obtained from wild or planted groves of acacia that are tapped in the dry season (October) by stripping an approximately 2-ft by 2-in. (60 by 5 cm) section of bark and hand-collecting the dried tears (weighing 20–200 g) over a four- to eight-week period. Total production is 60,000–80,000 tons, with about one-half being exported to the United States. The government controls production and price. Although the trees are beneficial for the soil and are often planted to prevent desert encroachment, gum harvesting is a laborious and financially unrewarding task. As with all exudate gums, increases in labor costs reduce harvests and increase gum cost.

Structure

Gum arabic is a neutral or slightly acidic (pH 4.5–5.0) gum containing calcium, magnesium, potassium, and at times, other cations. A typical analysis shows: ash, 3%; nitrogen, 0.29%; methoxyl, 0.25%; specific rotation, $-30°$; intrinsic viscosity 13.4; uronic acid, 16%. Generally accepted values for number-average and weight-average molecular weights are 250,000 and 580,000, respectively; these values correspond to degrees of polymerization of 1,550 and 3,600 and indicate polydispersity. The equivalent weight is approximately 1,100. Hydrolysis of a typical gum preparation produces D-galactose, 44%; L-arabinose, 24%; D-glucuronic acid, 14.5%; L-rhamnose, 13%; and 4-*O*-methyl-D-glucuronic acid, 1.5%.

Gum arabic is a heterogeneous material, generally consisting of two fractions. One, which accounts for about 70% of the gum, is composed of polysaccharide chains with little or no nitrogenous material. The other fraction contains molecules of higher molecular weight that have protein as an integral part of their structures. The protein content of the protein-polysaccharide fraction varies. Polysaccharide structures are covalently attached to the protein component by linkage to hydroxyproline and, perhaps, serine units, the two predominant amino acids in the polypeptide. The overall protein content is about 2 wt%, but specific fractions may contain as much as 25 wt% protein.

The polysaccharide structures, both those attached to protein and those not, are highly branched, acidic arabinogalactans. They contain main chains of (1→3)-linked β-D-galactopyranosyl units having two- to four-unit side chains also consisting of (1→3)-linked β-D-galactopyranosyl units joined to it by (1→6) linkages. Both the main chain and the numerous side chains have attached α-L-arabinofuranosyl, α-L-rhamnopyranosyl, β-D-glucuronopyranosyl, and 4-O-methyl-β-D-glucuronopyranosyl units. The two uronic acid units occur most often as terminal ends of the chain branches. A repeating unit with a molecular weight of about 8,000 (degree of polymerization about 50) has been suggested.

Determination of the hydrodynamic radius suggests a very compact structure, which explains why concentrations 20–40 times greater, compared to other gums, are required for equivalent viscosities. A model has been proposed in which, on average, each protein-polysaccharide molecule consists of five polysaccharide units attached to one protein unit. The polypeptide must be at the periphery of the composite molecule because it can be easily degraded by proteolytic enzymes, which probably explains gum arabic's emulsifying potential, that is, its surface activity.

Glycoproteins in which arabino-oligosaccharides are linked to hydroxyproline and D-galactose is attached to serine have also been detected in whole gum arabic.

Properties

Gum arabic dissolves easily when stirred in water. It is unique among the food gums because of its high solubility, the low viscosity of its solutions, and the Newtonian flow of its solutions at concentrations below about 40% concentration. Solutions of 50% concentration can be made. At this concentration, the dispersion is somewhat gel-like. Highest viscosities are produced at its normal solution pH of 6. As with other carboxylic-containing polysaccharides, its solubility is minimum below about pH 3, where ionization of the carboxyl groups is repressed.

Gum arabic is both a fair emulsifying agent and a very good stabilizer for emulsions of citrus and other flavor oils. For a gum to have an emulsion-stabilizing effect, it must have anchoring groups with a strong affinity for the surface of the oil and a molecular size large enough to cover particle surfaces. Gum arabic has surface activity and forms a thick, sterically stabilizing, macromolecular layer around flavor oil droplets.

Stabilized flavor oil-in-water emulsions prepared with gum arabic can be spray-dried. In addition to its encapsulating properties, high-solid, low-viscosity solutions can be made with gum arabic. Its heat stability permits spray-drying at higher temperatures than is possible when some substitutes are used. This effects a more rapid removal of water, which exposes the flavor to heat for a shorter time. Because gum arabic is nonhygroscopic, it is not necessary to dry the air to which the powder is exposed as it is cooled and conveyed out of the drier.

Another important characteristic is its compatibility with high concentrations of sugar. Therefore, it finds widespread use in confections that have high sugar content and low water content, where it retards or prevents crystallization of sugar.

Uses

More than half the world's supply of gum arabic is used in confections such as caramels, toffees, jujubes, and pastilles. These and other typical products are listed in Table 14.1. In confections, gum arabic prevents sucrose crystallization and emulsifies and distributes fatty components so as to retard their accumulation on the surface. Avoidance of lipid migration to the surface is important because such migration results in a greasy whitening, called *bloom*. Another use of gum arabic is as a component of the glaze or coating of pan-coated candies.

Another broad use of gum arabic is emulsification of flavor oils. It is the gum of choice for emulsification of citrus, other essential oils, and imitation flavors used as baker's emulsions and concentrates for soft drinks. Gum arabic has the unique ability to stabilize flavor oil-in-water emulsions, both as concentrates (20% oil by weight) and highly diluted in beverages. The soft drink industry

TABLE 14.1
Typical Products Containing Gum Arabic

Caramels	Liquid diet products
Cocktail mixes	Lozenges
Dry soup mixes	Pan-coated candies
Filling for chocolates	Pastilles
Fruit drink mixes	Pickle relish
Fruit juices	Soft drinks
Gum drops	Toffees
Jujubes	

consumes about 30% of the gum supply for use as an emulsifier and stabilizer.

Emulsions made with flavor oils and gum arabic can be spray-dried to produce dry, flavor powders that are nonhygroscopic and in which the flavor oil is protected from oxidation and volatization. Rapid dispersion and release of flavor without affecting product viscosity are other attributes. These stable flavor powders are used in dry package products such as beverage, cake, dessert, pudding, and soup mixes.

Free-flowing powders made by spray-drying an emulsion of gum arabic and an oil are used as clouding agents in dry drink mixes.

Gum arabic once had many more applications, which have become minor. It has been replaced by pectins and various thin-boiling starch products with similar properties and greater stability of supply and price. It is still used to some extent in the preparation of gum drops, cough drops, and other lozenges. It provides gloss and flexibility when used as a bun glaze, and its emulsifying action can be employed in toppings, icings, spreads, frozen desserts, and whipped or aerated, reduced-calorie margarines.

Gum arabic may be labeled *gum arabic, arabic gum, gum acacia*, or *acacia gum*.

Gum Karaya

Gum karaya is an acetylated polysaccharide exudate of *Sterculia urens* trees that grow in the dry, rocky hills and plateaus of central and northern India, often near the areas of production of gum ghatti. Collection is in April to June and again in September after the monsoons. Trees are tapped by removing bark and collecting the dried exudate. About 80% of the annual production of 5,000–6,000 tons is exported to the United States.

Gum karaya and *karaya gum* are acceptable label designations.

Gum Ghatti

Gum ghatti is exuded from the *Anogeisusus latifolia*, family Compretaceae, tree found mainly in the dry deciduous forests of India and to a lesser extent in Sri Lanka. It occurs in tears of about 2–5 in. (5–13 cm) in diameter but most often in large vermiform masses. Some 1,200 tons are collected annually.

Gum ghatti and *ghatti gum* may be used on labels.

Gum Tragacanth

Gum tragacanth is another ancient gum of commerce, being described by Theophrastus in the 3rd century B.C. It is the exudate of the low-growing *Astragalus* bush, which has a long, deep root that is tapped. The exudate is in the form of a long, curving string or ribbon of rapidly hardening gum. The name *tragacanth* is derived from the Greek *tragos* meaning goat and *akantha* meaning horn, which describes the appearance of the exudate. The leguminous plants are common in Iran, Syria, and Turkey. Gum collections have been small in recent years, mainly because of labor costs. Although at one time U.S. requirements were almost 1,000 tons per year, prices rose to the extent that the food industry sought substitutes. Experimental growth of Iranian plants adapted to Arizona and California is under way, but the cost of production remains high and little gum tragacanth is used.

Gum tragacanth may still be found in pourable salad dressings, pickle relish, bakery products, and products such as frozen gravy and frozen cheese lasagna.

Gum tragacanth and *tragacanth gum* may be used on labels.

Chapter 15

Bulking Agents, Fat Mimetics, and Carbohydrate Nutrition

In addition to providing texture, carbohydrates are the principal providers of bulk in food products. The higher saccharides may be digestible, partially digestible, or nondigestible. When digestion occurs, the products of digestion are absorbed and catabolized. Those escaping the small intestine may be metabolized by microorganisms in the large intestine, where they produce substances that are absorbed and catabolized, providing some energy. Therefore, carbohydrates may be caloric, partially caloric, or essentially noncaloric. They may be soluble or insoluble, and they may produce high or low viscosities. In addition to possessing the physical property of bulk, certain carbohydrates, in aqueous dispersion, may induce other beneficial sensory perceptions, such as sensations of thickness, creaminess, pulpiness, or chewiness—even the perception of fattiness. Naturally occurring plant carbohydrates are nontoxic. Since before recorded history, they have been the principal source of metabolic energy for humans and the means for maintaining the health of the human gastrointestinal tract.

Bulking Agents

Natural food gets most of its bulk from remnants of plant cells resistant to hydrolysis by enzymes of the alimentary tract. Cellulose, hemicelluloses, pectin, and lignin are components of these remnants. This nondigestible material is classified as "dietary fiber," although dietary fiber is not necessarily fibrous in nature. *Dietary fiber* is a nutritional term that has nothing to do with its physical or chemical nature, although both chemical and physical properties are involved in its determination. It is actually defined by the method used to measure it, of which there are several. Both insoluble plant cell-wall materials, primarily cellulose and lignin, and nonstarch, water-soluble polysaccharides are components of dietary fiber. Even starch materials that escape digestion (called *resistant starch*) are classified as dietary fiber. The only common feature of these substances is that they are nondigestible polymers. The original definition was based on early methods for analysis that used partial hydrolysis with acids, allowing recovery of the fibrous-appearing cell-wall residue. New analytical methods use amylolytic enzymes and permit recovery of all nonstarch polysaccharides and any resistant starch. Therefore, not only do natural components of food contribute "dietary fiber," so do gums that are added to modify rheological properties, to provide bulk, and/or to provide other functionalities as previously described.

Low-molecular-weight bulking agents that dissolve rapidly are used to provide weight and volume for coffee-whiteners and sometimes for packaging high-intensity sweeteners. Such bulking agents are mainly low-molecular-weight, bland-tasting, starch dextrins. Dextrin-like polymers prepared from D-glucose by polymerization with an acidic catalyst are available commercially and used under the ingredient designation *polydextrose*.

Dietary fiber bulking agents are important in nutrition because they maintain normal functioning of the gastrointestinal tract. They increase intestinal and fecal bulk, which lowers intestinal transit time and helps prevent constipation. Their presence in foods induces satiety at mealtime. Vegetarians routinely consume 40–50 g of dietary fiber daily with beneficial effects. Nutritional experts set requirements of dietary fiber at 25–50 g per day. Fiber bulking agents are claimed to decrease blood cholesterol levels, lessening the chance of heart disease, and reduce the chances of colonic cancer, probably due to their sweeping action.

One natural component of dietary fiber is β-glucan (Chapter 8). β-Glucans are nondigestible dietary polysaccharides that have gained prominence for lowering blood cholesterol. They are found in the bran of cereal grains and occur in especially high concentrations in oat and barley brans. They are linear polysaccharides composed of β-D-glucopyranosyl units joined in groups consisting mainly of cellotriosyl and cellotetraosyl units linked (1→3). A few chain segments have as many as four to eight consecutive units with (1→4)-linked sugar units. About 70% of the chain links are the (1→4) type.

When taken orally in foods, β-glucans reduce postprandial serum glucose levels and the insulin response (that is, they moderate the glycemic response) in both normal and diabetic human subjects. This effect seems to be correlated with viscosity. They also reduce serum cholesterol concentrations in rats, chicks, and humans. These physiological effects are typical of those of soluble dietary fiber. Gums other than β-glucans have similar effects in the gastrointestinal tract and on the amount of cholesterol in blood, but to different extents. Some that have been specifically examined are pectin (Chapter 13), guar gum (Chapter 9), xanthan (Chapter 10), and hemicelluloses (Chapter 8). For example, guar gum ingested at a rate of 5 g/day results in an improved glycemic index, a lowering of serum cholesterol, and no decrease in the high-density lipoprotein fraction that is the carrier of beneficial cholesterol. In addition to cereal brans, kidney and navy beans are especially good sources of dietary fiber.

Psyllium seed hulls are the source of psyllium gum, which can also be labeled *gum psyllium*. A product based on psyllium seed hulls has high water-binding properties, leading to rapid transit time in the gastrointestinal tract, and is widely sold to prevent constipation under the trade name Metamucil. A product with a methylcellulose base (Chapter 7) is sold for the same purpose under the trade name Citrucel.

Fat Mimetics

Governmental and other health organizations recognize that obesity increases the risk of cardiovascular problems and heart disease. Americans consume 138 lb (62.6 kg) of fat per person per year; in other words, the average American receives about 38% of his/her calories from fat, although health experts recommend that fats not contribute more than 30% of calories. Dietary fats are especially responsible for obesity because they are the most concentrated source of energy, contributing 9 kcal per gram, compared to the 4 kcal per gram contributed by digestible carbohydrates.

Since people are reluctant to give up the sensation of fattiness, a search for fat replacers is under way. An ideal fat replacer would have all the functional characteristics of fat, but with no or significantly fewer calories. The functional properties of fats that must be mimicked include organoleptic properties, lubricity, flavor, aroma, heat stability, color, spreadability, emulsification, and aeration. The ideal fat replacer does not exist; rather, a variety of products that produce the perception of fat are available as fat-sparing ingredients.[1] These products are based on either carbohydrates, proteins, or fats, with the largest number in use at this time being carbohydrate-based.

The fatty sensation is extensively dependent on rheology and less on the nature of the ingredient. Specially modified carbohydrates and proteins have been developed to provide the required rheology.

The first fat mimetic marketed was composed of whey protein microspheres, 0.1–2 μm in diameter, combined with an antiaggregating agent such as xanthan. A similar product is made from egg white protein. Fat mimetics that produce a mild fatty sensation are also made from carbohydrates. In each instance, the mimetic consists of microparticles (in some cases, microcrystals).

Fat mimetics can be made from a variety of starches. One mimetic is a maltodextrin made by partial enzyme-catalyzed hydrolysis of potato starch and is recommended to replace up to 50% of fat in food products.[1] Another is a tapioca maltodextrin derivatized to low degree of substitution levels with an alkenylsuccinate. When 0.5% of a 25% suspension of the latter is added to ice cream containing the required minimum of 10% butterfat, the ice cream has the eating quality of one containing 14% butterfat. Another fat mimetic is made by treating granular corn starch with dilute acid or with α-amylase to a dextrose equivalence of less than 5, preferably about 2 (see Chapter 6). In both cases, the amorphous regions of the granule are preferentially removed by hydrolytic erosion, leading to a more crystalline product. In another method, starch is spray-dried through a special nozzle designed to receive a starch-water slurry and steam in such a manner that the starch granules are swollen, but not extensively disrupted. The product consists of particles of such size that 90% will pass a 100-mesh screen. Granules may be mildly crosslinked with about 0.01% phosphoryl chloride to provide additional stability; crosslinking is especially needed with tapioca and other tuber and root starches. Small-granule starches such as those from amaranth, which are about 1 μm in diameter, have some fatty sensory character in slurry, particularly

---

[1] Since it is generally recommended that fat mimetics replace only a part of the fat in a product, they are often called *fat sparers*, sometimes *fat replacers*.

after mild glucoamylase treatment, and are the subject of current research.

No one fat mimetic, as yet, provides all the attributes of fat, and a matrix of materials has been the best approach.

Consumers demand not only reduced-calorie foods, but also the taste of fat. The fat mimetics described above do not melt like fats, nor do they carry flavors like fats. Therefore, research efforts have been, and are being, also directed toward development of lipidlike fat substitutes. Many of these are fatty acid polyesters of carbohydrates.[2]

Polyesters for use as fat substitutes have been prepared from fatty acids and sucrose, raffinose, stachyose (Chaper3), trehalose, sorbitol (Chapter 2), and alkyl glycosides (Chapter 1). The only one that is a commercial product and approved for (specific, limited) use in foods is the product sold under the trade name Olestra (Chapter 3). It is a sucrose polyester containing six to eight fatty acyl groups per sucrose molecule. All such compounds are lipophilic, nondigestible, and nonabsorbable, providing the chemical and physical properties of conventional fats and oils without contributing caloric value.[3]

Carbohydrate Nutrition

Carbohydrates are generally innocuous edible components of foods. They have always been the principal source of energy and the means for maintaining health of the gastrointestinal tract for humans as a whole. The digestible carbohydrates in an average human diet in the United States include the groups shown in Table 15.1.

Only monosaccharides can be absorbed from the digestive tract, so only monosaccharides (D-glucose and D-fructose are the only significant ones in the human diet) need not undergo hydrolysis before absorption. All other carbohydrates must be digested before their monomeric units can be absorbed and catabolized to provide energy. The sucrase of the small intestine catalyzes the hydrolysis of sucrose into D-glucose and D-fructose, and lactase converts lactose

---

[2] Conventional fats and oils are also fatty acid polyesters of a carbohydrate, that is, fatty acid peresters (triesters) of an alditol, glycerol.

[3] Fatty acid esters of sucrose, sorbitol, and other carbohydrates with lower degrees of substitution are at least partially digestible and are used in foods as emulsifiers (Chapter 3).

into D-galactose and D-glucose. The only other carbohydrases acting in the small intestine, where digestion and absorption take place, are pancreatic α-amylase and intestinal maltase and isomaltase. The three enzymes together convert the starch polysaccharides and starch-related products into D-glucose, so the starch polysaccharides are the only polysaccharides that can be hydrolyzed by human digestive enzymes. The D-glucose produced by this digestion is absorbed by microvilli of the small intestine to supply metabolic energy.

Polysaccharides may be digestible (most starch-based products), partially digestible (retrograded amylose, the so-called "resistant starch"), or nondigestible (essentially all other polysaccharides). When digestive hydrolysis to monosaccharides occurs, the products of digestion are absorbed and catabolized for energy. Therefore, carbohydrates may be caloric, partially caloric, or essentially noncaloric. Other polysaccharides consumed normally as natural components of edible vegetables, fruits, and other plant materials and those food gums added to prepared food products are not digested in the upper digestive tract of humans but pass into the large intestine with little or no change, for no enzymes exist in the human gastrointestinal system for hydrolyzing polysaccharides other than starch.[4] Furthermore, the acidity of the stomach is neither strong enough, nor is the residence time of polysaccharides in the stomach sufficiently long, to cause significant chemical cleavage.

However, when undigested polysaccharides reach the large intestine, they come into contact with the normal intestinal microor-

---

[4] As mentioned, only monosaccharides are absorbed and end up in the blood stream. Also, as mentioned, the only other nonmonosaccharide dietary components that are digested, that is, hydrolyzed by human enzymes, are the disaccharides sucrose and lactose and the starch polysaccharides, so generally only D-glucose, D-fructose, and D-galactose are used as energy sources.

TABLE 15.1
Digestible Carbohydrates in the Average U.S. Diet

Digestible Carbohydrates in Diet	Amount
Starch (polysaccharide)	About 60%
Sucrose (disaccharide)	About 30%
Lactose (disaccharide)	About 5%
Glucose and other monosaccharides	About 5%

ganisms, some of which produce enzymes that catalyze hydrolysis of certain polysaccharides or certain parts of polysaccharide molecules. The consequence of this is that some polysaccharides not cleaved in the upper intestinal tract may undergo cleavage and microbial metabolism within the large intestine. When this occurs, the molecular weight of the polysaccharides is reduced.

Sugars that are split from the polysaccharide chain are used by the microorganisms of the large intestine as energy sources in anaerobic fermentation pathways that produce lactic acid and volatile fatty acids, such as propionic, butyric, and valeric acids. These short-chain acids can be absorbed through the intestinal wall and metabolized, primarily in the liver. In addition, a small, although significant in some cases, fraction of the released sugars can be taken up by the intestinal wall and transported in the portal blood stream to the liver and metabolized. It is calculated that, on average, 7% of human energy is derived from sugars split from polysaccharides by microorganisms in the large intestine and/or from the acid by-products produced from them by these microorganisms through anaerobic fermentation. The extent of polysaccharide cleavage depends on the abundance of the particular organism producing the specific enzymes required. Thus, when changes occur in the type of polysaccharide consumed, utilization of the polysaccharide may temporarily be minimal until there is an increase in the population of colonic microorganisms that produce the enzymes required for splitting the new polysaccharide and/or the microorganisms adapt by producing the required enzymes.

Some polysaccharides survive almost intact during their transit through the entire gastrointestinal tract. These, plus the larger segments of other polysaccharides, give bulk to the intestinal contents and lower transit time. They can be a positive factor in health through a lowering of blood cholesterol, perhaps by sweeping out bile salts and reducing their chances for reabsorption from the intestine. In addition, the presence of large amounts of hydrophilic molecules maintains a water content of the intestinal contents that results in stool softness and consequent easier passage through the large intestine. Lignin is not degraded and passes through the digestive tract essentially unchanged. If large quantities of new, slowly hydrolyzed polysaccharides are present in the diet, they will tend to decrease the normal 25-hr average transit time and, also, sweep out of the intestine significant quantities of the bacteria normally present at about a 20% dry weight level. Then, the microbial population becomes reconstituted by increasing the existing population, with an increase in the number of organisms that can more readily hydrolyze the new polysaccharide introduced into the diet.

Health organizations recommend that 55–60% of the calories consumed be carbohydrate. This nearly equates to the amount of starch in the diet. However, not all starch is equally digestible. A portion of starch in starch-based foods escapes digestion in the small intestine, and starch can be found in feces at times. Factors that limit starch digestion include degree of gelatinization, granule size, amylose content, starch-protein interactions, starch-lipid complexes and, perhaps most importantly, the degree of crystallinity, including that formed by retrogradation during processing. Freshly cooked starch is most readily digested. Starch resistant to digestion passes to the large intestine, where it may be fermented by the colonic microflora. Freeze-thaw cycling increases retrogradation, increasing resistance to digestion. Although low, starch's glycemic index (that is, the blood glucose elevation compared to a standard, which is normally D-glucose) is further depressed in starch with a high level of crystallinity resulting either from high amylose content, from treatments with heat and moisture, from freeze-thaw cycling, or from other processes. High-amylose starch causes only a mild rise in the glycemic index. Under normal cooking, high-amylose starch becomes even more resistant to digestion.

Water-soluble polysaccharides, such as guaran, ingested at the rate of 5 g/day, show a glycemic index of fasting with a 13% lowering of serum cholesterol and no decrease in high-density lipoproteins, the beneficial cholesterol carrier.

Many experimental investigations have been conducted in which the fate of polysaccharides has been followed in the gastrointestinal tract of humans and animals, and there has never been observed an instance of acute toxicity from oral intake of reasonable levels of natural polysaccharides. Present evidence indicates that chemical and biochemical modification of high-molecular-weight polysaccharides is likewise innocuous because the modified polysaccharides (other than starch) are nondigestible dietary components.

Chapter 16

Sweeteners

Much work has been done on the theory of sweetness, the structural requirements of molecules that enable them to impart sweet taste, mapping of the various sweet taste receptors, and the relationship of sweetness to other flavors.

Nutritive Sweeteners

Nutritive sweeteners are substances that impart a sweet taste and provide calories (energy). They are sucrose, D-fructose, D-glucose (dextrose), syrups (corn syrups, high-fructose corn syrups [HFCS], maple syrup), honey, molasses, sugar alcohols (D-glucitol, mannitol, hydrogenated syrups), maltodextrins, neosugar, and aspartame. Aspartame, a dipeptide ester (see Nonnutritive Sweeteners), is categorized as a nutritive sweetener by the U.S. Food and Drug Administration (FDA) because, on an equal-weight basis, it has approximately the same caloric content as a carbohydrate. However, since about 200 g of sucrose can be replaced by 1 g of aspartame because of the greater sweetness potency of aspartame, it effectively functions as a nonnutritive sweetener.

Nutritive sweeteners have functional properties other than providing sweetness. They also provide body (viscosity) to liquids, bulk, and desirable texture and mouthfeel. They bind and hold water, keeping some products moist and preserving and extending the shelf life of others by reducing water activity. They lower the freezing points of ice cream and frozen desserts; generate crust color and flavor via Maillard and/or nonenzymic browning reactions (Chapter 2) in bakery products; and are a source of carbon for fermentation, for example, in the manufacture of bread, other yeast-leavened baked goods, pickles, and alcoholic beverages. They offset bitter or sour tastes and enhance and preserve flavors, the latter because they possess antioxidant properties.

Sucrose and other nutritive carbohydrate sweeteners that are contributors to the caloric content of the human diet have been presented in Chapter 3 (sucrose, oligosaccharides related to sucrose, lactose), Chapter 6 (glucose/dextrose, fructose, corn syrups, HFCS, maltodextrins), and Chapter 3 (isomaltulose, leucrose, neosugar). The polyols (alditols, Chapter 2) are considered to be nutritive sweeteners, but they have lower caloric values because they are more slowly and poorly absorbed from the digestive tract.

Sweetness is application-dependent, and relative sweetness values depend on a number of factors, primarily temperature and concentration. Sweetness is also influenced by the solids content, viscosity, pH, and the concentration of other sweeteners and flavors. However, relative sweetness values of nutritive sweeteners, compared to sucrose as the standard, are basically as shown in Table 16.1.

Sweetness responses also vary with time (Fig. 16.1). Therefore, in food and confectionery products, both the intensity and duration of sweetness can be controlled by using mixtures of sweeteners. Different sweeteners also have different effects on other ingredients,

TABLE 16.1
Relative Sweetness Levels of Nutritive Sweeteners

Carbohydrate	Relative Sweetness
β-D-Fructopyranose	160–180
90% HFCS[a]	120–160
D-Fructose	140
55% HFCS	>100
Invert sugar	>100
Sucrose	100
42% HFCS	100
Xylitol	100
D-Glucose	70–80
70-DE[b] Corn syrup	70–75
Regular corn syrup	~50
D-Glucitol (Sorbitol)	50
Maltose	30–50
Lactose	20

[a] High-fructose corn syrup. The number indicates the percent of carbohydrate that is D-fructose, the remainder being D-glucose.
[b] Dextrose equivalent (see Chapter 6).

modifying, for example, the gelatinization temperatures and the viscosities and gel strengths of products containing starch (Chapter 6). Both nutritive and nonnutritive sweeteners vary in taste quality, the flavor profile being most meaningful in the context of the food product in which it is used. Sweeteners may account for from 25 to 90% of the weight of confectionery products.

A task force from the FDA Center for Food Safety and Applied Nutrition examined nutritive carbohydrate sweeteners with respect to their content in food, consumer intake of them, and trends in both these areas. In a report that included approximately 1,000 references (W. H. Glinsmann, H. Irausquin, and Y. K. Park, Evaluation of Health Aspects of Sugars Contained in Carbohydrate Sweeteners. Report of Sugars Task Force, 1986. *J. Nutrition* 116:S1-S216, 1986), the task force concluded that the amount of added sugars in the average American's diet (about 11% of total caloric intake) does not contribute to any health problems, other than dental caries. The report concludes that sucrose, invert sugar, and corn syrups, including HFCS, as they are commonly consumed in the American diet, do not cause glucose intolerance, diabetes mellitus, problems with blood lipids, cardiovascular disease (including hypertension and atherosclerosis), behavioral problems, obesity, calciuria, gall-

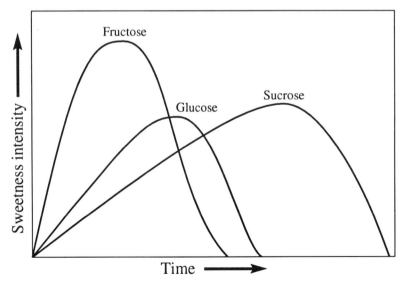

Fig. 16.1. The relative time-dependent sweetness responses of sucrose, D-fructose, and D-glucose

stones, reduced bioavailability of other nutrients, or cancer. Other national and international organizations reviewing the available data before and since 1986 have come to identical conclusions. At least 20 controlled scientific studies have been done specifically to determine whether there is any relationship between sugar and hyperactivity (attention deficit disorder) or any other behavior of children in particular. The conclusion reached from these studies, which appear to have been repeated over and over because of a belief that earlier studies used an improper experimental design or must have missed something, is that there is no evidence for a cause-and-effect relationship between sugar and behavior. Sugars have also been cleared of a role in kidney failure, cataracts, increased risk of disease, and reduced life expectancy.

Nonnutritive Sweeteners

A nonnutritive sweetener is, by definition, any substance that imparts a sweet taste, is not catabolized by the human body, and therefore does not provide calories (energy). Aspartame and other peptide sweeteners are digested and catabolized, but because they are high-potency sweeteners and therefore used in only small amounts, they do not provide appreciable calories to diets and are usually considered to be nonnutritive.

Some products used to sweeten food products labeled as "lite" are nutritive. They are products such as 90% HFCS, which can be used to make products, such as pancake syrups. The syrups are labeled "lite" because the greater sweetness of the 90% HFCS allows the total carbohydrate to be reduced to as little as one-half the normal amount.

Nonnutritive sweeteners may be used in diets requiring calorie restriction, although the calorie contribution by nutritive sweeteners is almost always only a fraction of that contributed by fats. They may be most important in the reduction or prevention of dental caries. Use of xylitol (Chapter 2) is important in this regard.

The principal nonnutritive sweeteners and their approximate relative sweetness values are given in Table 16.2.

Aspartame

Aspartame is the nonnutritive sweetener used in greatest amounts. It is a dipeptide methyl ester, specifically L-aspartyl-L-phenylalanine methyl ester. It has a sucroselike taste, with a relatively slow sweetness onset and a sweet aftertaste. It acts synergistically with saccharin, cyclamate, and malic acid and enhances and extends citrus fruit flavors. Its maximum stability is in the pH range 3–5, which presents no problem for most food and beverage products. However, with time, in products such as soft drinks, acid-catalyzed hydrolysis cleaves the methyl ester group, leading to loss of sweetness. Heating at pH values above 5 results in the conversion of some aspartame molecules to a diketopiperazine, also resulting in loss of sweetness. A granulated, encapsulated form, in which the protective coating acts as a time/temperature release system, is available for use in bakery products.

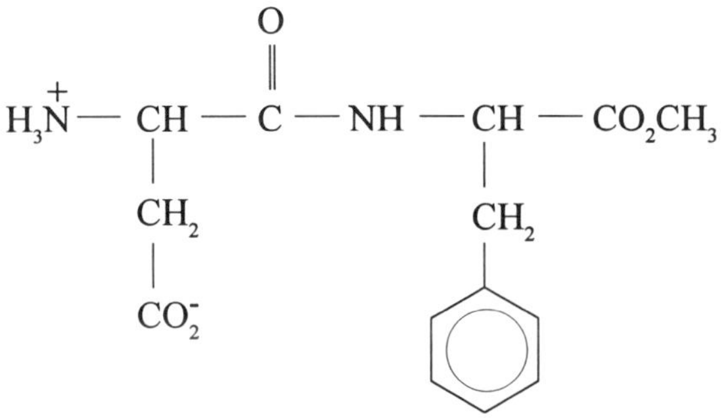

$$H_3\overset{+}{N}-CH(CH_2CO_2^-)-C(=O)-NH-CH(CH_2C_6H_5)-CO_2CH_3$$

Aspartame

Aspartame has been found to be safe, as could be predicted because it is the methyl ester of a dipeptide of two natural amino acids. However, because it contains L-phenylalanine, which is harmful to people with the disease phenylketonuria, products containing

TABLE 16.2
Relative Sweetness Levels[a] of the Principal Nonnutritive Sweeteners

Compound	Relative Sweetness
Alitame	200,000–290,000
Sucralose	55,000–75,000
Saccharin	30,000
Aspartame	18,000–20,000
Acesulfame K	15,000–20,000
Cyclamate	3,000

[a] Sucrose = 100.

it are required to contain the statement "Phenylketonurics: contains phenylalanine."

Saccharin

Saccharin contains the -CONHSO$_2$- (*N*-sulfonyl amide) structural unit common to several compounds with a sweet taste. Of the three nonnutritive sweeteners approved for use in the United States, saccharin has the greatest relative sweetness. It provides a rapid sweetness onset. For about one-third of the human population, the sweet taste of saccharin is accompanied by significant metallic and bitter aftertastes; the remaining two-thirds perceives these off-tastes to degrees that range from moderate to zero. Saccharin is stable and, of all the high-potency sweeteners, has the widest range of applications. It acts synergistically with aspartame and/or cyclamate. It is not metabolized by humans. Saccharin is used in more than 80 countries, but it is banned in Canada as a carcinogen. In the United States, a congressionally mandated moratorium on regulatory action against it has been renewed every two years since 1977. Saccharin is the most economical of the sweeteners, having a "cost per sucrose equivalent (CSE)" of about $0.01/lb.

Saccharin

Acesulfame

Acesulfame, the potassium salt of which is known as acesulfame-K, also contains an *N*-sulfonyl amide unit. Like saccharin, it is heat stable, pH tolerant, and synergistic with other high-potency sweeteners. It is FDA approved. Like saccharin, some perceive it as having a metallic and bitter aftertaste, the population also being heterogeneous with respect to sensitivity to this characteristic. It provides a quick onset of sweetness. Acesulfame is the only high-potency sweetener now on the U.S. market that does not require a warning label.

Others

Aspartame, saccharin, and acesulfame are the only high-potency sweeteners approved for use in the United States, but others are under consideration. Sucralose (Chapter 3) has a clean sweetness, with the same flavor profile as sucrose and a relative sweetness about 600 times that of sucrose. It has good solubility and heat and acid stability. It has been approved for use in Canada since 1991.

Another dipeptide analog, alitame, has a relative sweetness at least 2,000 times that of sucrose. It has a clean flavor profile like that of aspartame, but it is more stable to hydrolysis, that is, to heat and pH in the range 5–8. It acts synergistically with acesulfame-K and/or cyclamate.

Cyclamate was the most widely used artificial sweetener in the 1960s. Its use in the United States has been prohibited since 1970, but it is approved for use in more than 40 countries, including Canada. It contains the -NH-SO$_3$- (sulfamate) structural unit common to several compounds with a sweet taste. Sold as the sodium or calcium salt, cyclamate is soluble and stable, has a sugarlike taste, and acts synergistically with saccharin and/or aspartame. It is partially absorbed but not catabolized.

$$\text{C}_6\text{H}_{11}\text{—NH-SO}_3^-$$

Cyclamate

Each nonnutritive, low-calorie sweetener performs better in some products than in others. From 1910 (when it was first used commercially) until the 1960s, saccharin was the only nonnutritive sweetener available. Cyclamate was introduced in the 1950s. In the 1960s, these two sweeteners were often used together in diet soft drinks, tabletop sweeteners, and other low-calorie food and beverage products. Saccharin boosts the sweetening power of the less potent cyclamate; cyclamate eliminates most of the aftertaste of saccharin. Currently, a 10:1 mixture of cyclamate to saccharin is used widely in Europe.

Index

Acacia gum, 211–215
Acesulfame, 230, 231
Acetals, 11, 16, 45
   cyclic, 36–37
Acyclic structure, 3
Agar, 34, 188
   interaction with locust bean gum, 174
   uses, 194
Agar-agar, 194
Aglycon, 17, 43
Alcohols, polyhydroxy, *see* Alditols
Aldaric acids, 27–28
Aldehyde group
   oxidation of, 19–23
   reduction of, 23–27
Alditols, 23–27, 226
   acetates, 68
   methylated, acetates, 68–70
   percetates, 31
Aldobiouronic acid, 29–30
Aldohexose, 5
Aldonic acids, 19–23
Aldose, 3, 5–6
   oxidation of, 19–23
   reduction of, 23–27
Algin, 202
Algin derivative, 202
Alginates, 159
   ammonium, 202
   applications of, 199–202
   calcium, 198–201
   gelation, 198–199
   label designations, 202
   pH effects, 199
   potassium, 202
   preparation, 195
   properties, 198–199
   propylene glycol, *see* Propylene glycol alginate
   sodium, 195, 202
   sources, 195
   structure, 195–197
Alginic acid, 195, 199, 202. *See also* Alginates
Alitame, 231
Alkali cellulose, 160
Alpha-amylase, 139
Amadori rearrangement, 37–38, 40
Amylase
   alpha-, 139
   beta-, 27, 46–47, 140
   gluco-, 139–140, 221
Amyloglucosidase, 139–140, 221
Amylograph, *see* Brabender Visco/amylo/graph
Amylopectin, 81
   chains, A, B, and C, 119
   crystallinity of, 123–124
   potato starch, 121
   retrogradation of, 135
   structure, 119
Amylose, 80–81 complexes, 136–137
   leaching of, 128
   retrogradation of, 134
   structure, 122
3, 6-Anhydro ring, 34
Anogeisusus latifolia, 215
Anomeric carbon atom, 13
Anomeric configuration, 72
Anomers, 13
Arabic gum, 211–215
Arabinan, 65
Arabinogalactan, 65, 168–169

L-Arabinose, 7
Arabinoxylans, 65
Aspartame, 225, 228–231
Astragalus, 216
Axial positions, 13–15

Beet sugar, 54
Benedict reagent, 20
Beta-amylase, 27, 46–47, 140
Beta-glucan, 166–167, 218–219
Beta-glucanase, 167
Beta-limit dextrin, 140
Birefringence, 124–128
Boat conformation, 15
Brabender Visco/amylo/graph, 130–131, 147
Bran, 218
Brans, 165, 166
Brown sugar, 55
Browning, nonenzymic, 37–41
Bulking agents, 217–219

Cane sugar, 54–55
Caramel color, 40–41
Caramelization, 40–41
Carbohydrases, 222
Carbohydrate gum, 164
Carbohydrate nutrition, 221–224
Carbohydrate reactions, 19–41
Carbohydrates
    intestinal fermentation of, 223–224
    noncaloric, 217–224
    nondigestible, 217–224
Carbon atom
    anomeric, 13
    chiral, 2–4
Carboxyethylidene, 37
Carboxymethylcellulose, 34, 157
    degree of substitution, 160
    in ice cream, 193–194
    label designations, 161
    preparation, 160
    properties, 160–161
    structure, 160
Carob gum, *see* Locust bean gum
Carrageenans, 33–34, 187–194
    iota-type, 188–194
    kappa-type, 188–194
        in ice cream, 193–194
        interaction with locust bean gum, 172–174, 190, 193
    labeling, 194
    lambda-type, 188–194
    in meat products, 194
    Philippine natural grade, 187–188
    processed Euchema seaweed, 187–188
    reactivity with proteins, 192–194
Carrageenate, sodium, 187
Cassava starch, *see* Tapioca starch
Cellobiose, 45
Cellulose, 73, 78–79, 217
    acetate butyrate, 33
    applications of, 162
    bacterial, 156
    carboxymethyl-, *see* Carboxymethylcellulose
    crystallinity, 73–74, 153, 156–159
    diacetate, 33
    as dietary fiber, 154
    esters, 33
    ethers, 160–164
    gel, 159
    gum, 161
    hydroxypropyl-, 34
    hydroxypropylmethyl-, *see* Hydroxypropylmethylcelluloses
    label designations, 159
    methyl-, *see* Methylcelluloses
    microcrystalline, *see* Microcrystalline cellulose
    microfibrillated, 156
    microreticulated, 156
    powdered, 155
    sources, 154
    structure, 153
    triacetate, 33
    types, 161
Cellulose gel, 159
Cellulose gum, 161
Cellulosics, 153–164
Chiral, definition of, 2
Chitin, 73
Chocolate, 57
Cholesterol, 218, 219
Chondrus crispus, 187
Chondrus extract, 194
Citrucel, 219
Clathrates, 141
Cluster bean, 171
CMC, *see* Carboxymethylcellulose
Cohesiveness, 132
Confection sugar, 55

Conformations, 13–15
Corn fiber gum, 167–168
Corn starch
　equilibrium moisture, 128
　granule types, 124
　high-amylose, 121–122, 132
　properties, 121, 132
　waxy, 121, 132, 147–148
Corn syrup solids, 137
Corn syrups, 55, 137, 141–142, 226
Cosettes, 54
Crosslinked starch, *see* Starch, crosslinked
Cryoprotectants, 56, 80
Cyclamate, 231
Cyclic acetals, 36–37
Cycloamylases, 140–141
Cyclodextrin glucanotransferase, 140–141
Cyclodextrins, 140–141

Dairy products, carrageenans in, 192–194
Danish agar, 188
DE, *see* Dextrose equivalency
Debranching enzymes, 140
Decasaccharide, 43
Degree of polymerization, 63
Degree of substitution, 83
Dental caries, 26
Depolymerization, 84–86
Dextran, 26, 53–54
Dextrins, 137
Dextrose, 137–139
Dextrose equivalency, 137–138, 220, 222
Dietary fiber, 165, 167, 217–219
Differential scanning calorimetry, 129
Diheteroglycan, 64
Dinitrogen tetraoxide, 20
Disaccharide, 43, 46, 47, 51
DP, *see* Degree of polymerization
DS, *see* Degree of substitution
DSC, *see* Differential scanning calorimetry

Egg box, 198, 206
Emulsions, 213, 214–215
Enzyme-catalyzed hydrolysis, 86
Equatorial positions, 13–15
Esters
　acetate, 31, 33
　adipate, 33
　of cellulose, 33
　fatty acid, 33
　phosphate, 31–33
　of starch, 32–33
　succinate, 31–33
　sulfate, 33
Ethers, 33–36
　carboxymethyl, 33
　hydroxypropyl, 33
　methyl, 33
Eucheuma sp., 187

Fat mimetics, 210, 219–221
Fehling solution, 20
Fischer projection, 4
Flavor, 115
Fondant sugar, 55
Fringed micelles, 74
Frozen dessert, 192–194
D-Fructofuranosidase, beta-, 52
D-Fructose, 8, 25
　1,6-bisphosphate, 31–32
　production, 141–142
D-Fructose, 1-amino-1-deoxy-, 38
Furanose ring, 12
Furcellaran, 34, 188

G blocks, 196–197
Galactaric acid, 28
Galactomannan, 171–177
Galactose oxidase, 30
Galactosidase, beta-, 48
Galacturonans, 205
Galacturonoglycans, 203
GDL, *see* D-Glucono-delta-lactone
Gelatinization, 127–134
　temperature, 121–128
Gelation, 198, 112, 115, 163
　thermal, 163
Gellan, 65, 109–110
Gelling agent, choice of, 114–115
Gels, 107–114
　adhesiveness, 113
　alginate, 198
　brittleness, 112
　chewiness, 113
　cohesiveness, 113, 132
　egg box, 198
　elasticity, 113

formation, 81–81, 112, 163
fracturability, 112
gumminess, 113
hardness, 112
heat reversibility, 112, 115
properties, 81–82
shear reversibility, 112, 115
syneresis, 132
strength, 112
texture, 112–115
thermal reversibility, 112, 115, 163, 201, 208
thixotropic, 192, 199, 208–209
Ghatti gum, 215
Glass, 129
Glass transition temperature, 129
Glucan, beta-, 166–167, 218–219
Glucanase, beta-, 167
Glucaric acid, 28
D-Glucitol, 23, 34
anhydro-, see Sorbitan
dianhydro-, see Sorbitan
Glucoamylase, 139–140, 221
D-Gluconic acid
D-Glucono-delta-lactone, 22–23
Glucose syrups, see Corn syrups
D-Glucose, 3–6
6-phosphate, 31–32
production, 141–142
L-Glucose, 4
D-Glucosidase, alpha-, 52
Glucose oxidase, 21–23
D-Glucuronic acid
4-O-methyl-, 33
Glycan, 63. See also Polysaccharides
Glycaric acids, see Aldaric acids
Glycemic index, 219
Glyceraldehyde, 5, 7
Glycerose, 5, 7
Glyceryl monomyristate, 135
Glyceryl monopalmitate, 135
Glyceryl monostearate, 135
Glycitols, see Alditols
Glyconic acids, see Aldonic acids
Glycose, 9
Glycosides, 16–17, 45
Glycosulose, 9
Glycuronic acid, 29
Granules, starch
components of, 119–123, 126–127
damaged, 132

gelatinization of, 127–134
ghost, 129
glass transition of, 129
microscopy of, 118, 123–125
remnants, 129
structure, 123–124
types, 124–126
Guar gum
applications of, 175–176, 200
as dietary fiber, 219
in ice cream, 175, 193–194
interaction with microcrystalline cellulose, 159
interaction with xanthan, 173–174, 182
label designation, 177
methylation analysis, 34–35, 69–70
properties, 82, 172–174
source, 171
structure, 34–35, 69–70, 171
Guaran, 82, 171, 224
Gum acacia, 211–215
Gum arabic, 211–215
Gum ghatti, 215
Gum karaya, 215
Gum psyllium, see Psyllium seed gum
Gum tragacanth, 216
Gums, 75, 91
dissolution, 101–104
viscosity grades, 100–101

Hayworth projection, 11
Hemiacetal, 11
oxidation of, 21–23
Hemicelluloses, 165–169, 219
Heptasaccharide, 43
Heptose, 5, 6
Heptulose, 5, 9
Heteroglycans, 64, 74–75
Hexasaccharide, 43
Hexose, 5, 6
Hexulose, 5, 9
HFCS, see Syrups, corn, high-fructose
High-amylose corn starch, 122
properties, 121, 132
Hilum, 123–124
HMF, see Hydroxymethyl-2-furaldehyde
Homoglycans, 63, 73–74
HPMC, see Hydroxypropylmethylcelluloses

Humic substances, 41
Humin, 41
Hydrocolloids, 75, 91
Hydrogen peroxide, 30
Hydrolysis, 84–86
    acid-catalyzed, 137–139
    enzyme-catalyzed, 137–142
Hydroxyethyl, 83
Hydroxymethyl-2-furaldehyde, 38
H-4-Hydroxy-5-methylfuran-3-one, 41
Hydroxypropyl, 83
Hydroxypropylmethylcelluloses, 104, 158
    applications, 164
    label designations, 164
    molar substitution, 163
    properties, 163–164
    structure, 161
Hypobromite, 21
Hypochlorite, 21
Hysteresis, 99–100

Ice cream
    carboxymethylcellulose in, 193–194
    carrageenan in, 193–194
    guar gum in, 175, 193–194
    locust bean gum in, 175, 193–194
    microcrystalline cellulose in, 193–194
    xanthan in, 193–194
Inulin, 65
Invert sugar, 52
Invertase, 52
Irish moss extract, 194
Isoamylase, 140
Isomalt, 58–59
Isomaltitol, 58–59
Isomaltol, 41
Isomaltose, 226
Isomaltulose, 58–59
Isomerization, 9–11
Isomers, 9–11
Isosorbide, 36

Junction zone, 82, 172, 198, 206, 100, 107–114

Karaya gum, 215
Kestose, 60
Ketal, 11
Ketose, 5, 8–9
    reduction of, 25
Konjac mannan, 65

Lactase, 48–50
Lactitol, 50
Lactones, 22–23
Lactose, 30, 226
    digestibility, 222
    digestion, 47–50
Lactose intolerance, 49–50
Lactulose, 50
Larch, 168–169
Larix occidentalis, 168
LBG, *see* Locust bean gum
Leucrose, 58–59, 226
Lignin, 217
Limit dextrin, 140
Liquid sugar, 55
Locust bean gum
    applications, 172–177
    interaction with xanthan, 172–175
    label designation, 177
    methylation analysis, 34–35, 69–70
    properties, 82, 172–174
    source, 71
    structure, 171–172
Lysophosphatidylcholine, 126
Lysophospholipid, 126

M blocks, 196–197
Maillard browning, 37–40
Maillard reaction, 37–40, 57
Maize starch, *see* Corn starch
Maltitol, 27
Maltodextrins, 55, 137, 220, 226
Maltol, 41
Maltooligosaccharides, 138–141
Maltose, 27, 43–44, 46–47
Mannan, 73
    konjac, 65
D-Mannitol, 25–26
MC, *see* Methylcelluloses
MCC, *see* Microcrystalline cellulose
Meat products
    carrageenans in, 194
    PES/Philippine natural grade in, 194
Metamucil, 219
Methylation analysis, 34–35, 68–72
Methylcelluloses, 104, 219
    applications, 163–164
    degree of substitution, 162
    label designations, 164
    properties, 163–164
    structure, 161

Micelles, fringed, 74
Microcrystalline cellulose
  applications of, 159
  colloidal, 156–157
  in ice cream, 193–194
  powdered, 156
  preparation, 156–157
  properties, 154, 157–158
  types, 156–157
Milk, 47–50
Modified food starch, *see* Starch, modified food
Modified vegetable gum, 164
Modulus, 111
Molar substitution, 83
Molasses, 53–55
Monosaccharides, 1–17
  isomerization, 9–11
  ring forms, 11–17
Mouthfeel, 97–98
MS, *see* Molar substitution
Mucic acid, 28

Nata, 156
Nelson-Somogyi reagent, 20
Neosugar, 60–61, 226
Newtonian flow, 93–95
Nonasaccharide, 43
Nonenzymic browning, 37–40
Non-Newtonian flow, 94–100
Nonose, 5, 6
Nonreducing end, 43
Nonulose, 5, 9

Octasaccharide, 43
Octose, 5, 6
Octulose, 5, 9
Okra gum, 65
Olestra, 221
Oligomers, 43
Oligosaccharides, 43–61
  shorthand structures, 70–71
Oxidation, 19–23
  of aldehyde group, 19–23
  of hydroxyl groups, 27–30
  periodate, 28
Oxidation-elimination, 89

Palatinit, 58–59
Pastes, *see* Starch, pastes
Pasting, 127–134

Pectates, 203
Pectic acids, 203
Pectic substances, 203
Pectin, 217, 219
  acetyl groups in, 204
  amidated, 204, 209
  applications, 206–210
  DE, 204, 207
  degree of esterification, 204, 207
  depolymerization, 203
  as fat mimetic, 210
  gel properties, 208–210
  gelation of, 206–209
  gels, 158, 206–210
  high-methoxyl, 204–210
  HM, 204–210
  label designations, 210
  LM, 204–210
  low-methoxyl, 204–210
  methyl ester, 204
  nomenclature, 203–204, 208
  preparation, 203
  properties, 206–210
  sources, 203
  structures, 203–206
  types, 204–205
Pectinates, 203
Pectinic acids, 203
Pentasaccharide, 43
Pentose, 5, 6
Pentulose, 5, 9
Periodate, 28
PES, *see* Processed Euchema seaweed
PGA, *see* Propylene glycol alginate
Phytoglycogen, 121
Polarization cross, 124
Polydextrose, 218, 46
Polydisperse, 66
Polyhydroxy alcohols, *see* Alditols
Polymers, 43, 63
Polymolecular, 66
Polyols, 226. *See also* Alditols
Polysaccharides, 63–89
  acidic, 65, 79, 81
  bacterial, 66
  branched, 65–66, 77
  classification
    by source, 64
    by structure, 65
  conformations, 73–82
  crystallinity of, 73–74

depolymerization, 84–86
derivatization, 83–84
dissolution of, 73–76, 101–104
esters, 83–84
ethers, 83–84
gels, 107–114
hydration of, 73–82, 101–104
hydrolysis, 84–86
ionic, 65, 79, 81, 104, 106
linear, 65, 76–77, 81
methylation analysis, 34–35
modification, 83–89
molecular associations, 80–81
monomer constituents, 67
neutral, 55
oxidation-elimination, 89
preparation, 67–68
shorthand structures, 70–71
solution properties, 76–81, 104–107
solution rheology, 92–100
solution viscosity, 76–69, 92–100, 104–107
sources, 63–89
structural analysis, 68–72
synergisms, 107
temperature stability, 104–105, 114
viscosity, 104–107
viscosity grades, 100–101, 114
Potato starch
 equilibrium moisture, 128
 phosphate groups in, 126
 properties, 121, 126, 131–132
Powdered sugar, 55
Processed Euchema seaweed, 187–188
Propylene glycol alginate, 183–184
 gelation, 198–199
 interfacial activity, 199
 label designations, 202
 pH tolerance, 199
 preparation, 196–197
Protopectin, 65
Pseudoplastic flow, 93, 95–99
Pseudoplasticity, 181, 198
Psyllium seed gum, 65, 219
Pullulanase, 140
Pyranose ring, 11, 44

Raffinose, 54
Reactions, 19–41
Reducing end, 43
Reducing sugars, 20, 44

Reduction, 23–27
Resistant starch, 218, 222, 224
Retrogradation, 81, 131
Reversion, 45, 137
Rheology, 91–100
 pseudoplastic, 93, 95–99, 181, 198
 thixotropic, 99–100, 111
Rice starch, properties, 132
Ring
 boat forms, 15
 chair forms, 14–15
 conformations, 13–15
 furanose, 12
 pyranose, 11
 skew forms, 15
Ring forms, 11–17
Rosanoff method, 6
Rubber, 129

Saccharic acid, 28
Saccharin, 230, 231
Saccharose group, 5
Schardinger dextrins, 140
Seaweeds, red, 187–188
Septanose ring, 12
Setback, 131
Shorthand notations, 70–71
Skew conformation, 15
Sliminess, 97–98, 115
Sodium carboxymethylcellulose, *see* Carboxymethylcellulose
Sodium stearoyl 2-lactylate, 135
Solution properties
 of gums, 76–81
 of hydrocolloids, 76–81
 of polysaccharides, 76–81
Solution viscosity, 76–80
Sorbitan
 esters, 34–36
 ethoxylated esters, 36
 monolaurate, 36
 monooleate, 36
 monostearate, 34–36
Sorbitol, *see* D-Glucitol
SSL, *see* Sodium stearoyl 2-lactylate
Stabilization, 144
Stabilized starch, *see* Starch, stabilized
Stachyose, 54
Staling, 134–135
Starch, *see also* individual starches
 acid-modified, 137

240 / INDEX

amylopectin, *see* Amylopectin
amylose, *see* Amylose
  content of, 121
applications, 133
bleaching, 144
cold-water swelling, 149–150
cookup, 149
crosslinked, 143, 146–148
crystallinity, 123–124, 128
damaged, 132
derivatized, 144
dextrins, 137
digestibility, 222
enzymes, *see* Amylase
equilibrium moisture, 128
esters, 32–33, 144–145
ethers, 144–145, 148
gelatinization, 121, 127–134
gels, 158
granule properties, 121, 143
granule structure, 123–124
granules, 117–119, 123–131
helical nature, 122, 124, 136
hydrolysis, 137–142
hydroxypropyl, 34
instant, 148–149
lipids in, 121, 126–127
manufacture, 150–151
microscopy, 118, 123–125
minor components, 126–127
modified food, 142–150
octenylsuccinate, 145
oxidation, 144
paste properties, 121, 143
pastes, 94, 128
pasting, 127–134
phosphorus, 121, 126
potato, 32
pregelatinized, 148–149
protein content, 121, 127
relative viscosity, 121
resistant, 218, 222, 224
retrogradation, 131, 134
setback, 131
stabilized, 143–146
thin-boiling, 137
thinning, 137
waxy, 120
x-ray diffraction, 124–125
Sterculia urens, 215
Sucralose, 57–58, 231

Sucrase, 52
Sucrose, 51–59, 226
  commercial forms of, 55
  derivatives, 57
  digestibility, 222
  esters, 57
  fatty acid esters, 33
  octaacetate, 57
  oligosaccharides, 58
  properties, 55–57
  safety, 227, 228
  solution structure, 56
  sources, 53–54
Sugar beet, 54
Sugar cane, 53–54
Sugar
  beet, 54
  brown, 55
  cane, 54–55
  confection, 55
  fondant, 55
  invert, 52, 55
  liquid, 55
  nonreducing, 52
  powdered, 55
  reducing, 20, 44
  safety, 227–228
  soft, 55
  transformed, 55
Sulfate groups, 33
Surfactants, 136
Sweeteners, 225–231
  nonnutritive, 228–231
  nutritive, 225–228
Sweetness, 115
  relative, 226, 229
Syneresis, 82, 132, 143–144
Synergisms, 107
Syrups, 137
  corn, 225, 226
    high-fructose, 225, 226
Syrups, glucose, *see* Syrups, corn

Tapioca starch, properties, 121, 126, 131–132
Tetrasaccharide, 43
Tetrose, 5, 6
Tetrulose, 5, 9
Texture profile analysis, 111–113
Thermal gelation, 163
Thickening agent, choice of, 114–115

Thixotropic flow, 99–100, 111
Thixotropic gel, 192
Thixotropy, 199
Tollens silver mirror test, 19
Tragacanth gum, 216
Transformed sugar, 55
Tricalcium saccharate, 54
Triheteroglycan, 64
Triose, 5, 6
Trisaccharide, 43, 60, 69–70
Triulose, 5, 9

Uronic acid, 29

Visco/amylograph, *see* Brabender Visco/amylo/graph
Viscoelasticity, 91–92, 107–108
Viscosity, 76–79, 92–100, 104–107
Vitamin C, 23–25

Water
  activity, 39–40
  control of, 72–73
    by gums, 92–114
  freezable, 56
  nonfreezable, 56, 73, 80
  plasticizing, 73

Waxy starches, 120–121
  corn, *see* Waxy starches, maize
  maize, 121, 132, 147–148
Wheat starch
  granule types, 121
  properties, 121
Whey, 192–194
Wort, 167

Xanthan, 157, 219
  applications of, 182–185
  gum, 179, 185
  in ice cream, 193–194
  interaction with guar gum, 173, 174
  interaction with locust bean gum, 172–175
  preparation, 179
  properties, 181–185
  source, 179
  structure, 179–181
  synergisms, 182
Xanthan gum, *see* Xanthan
Xylan, 65, 165, 104, 168
Xylitol, 26
Xyloglucan, 65

Yield stress, 92–94